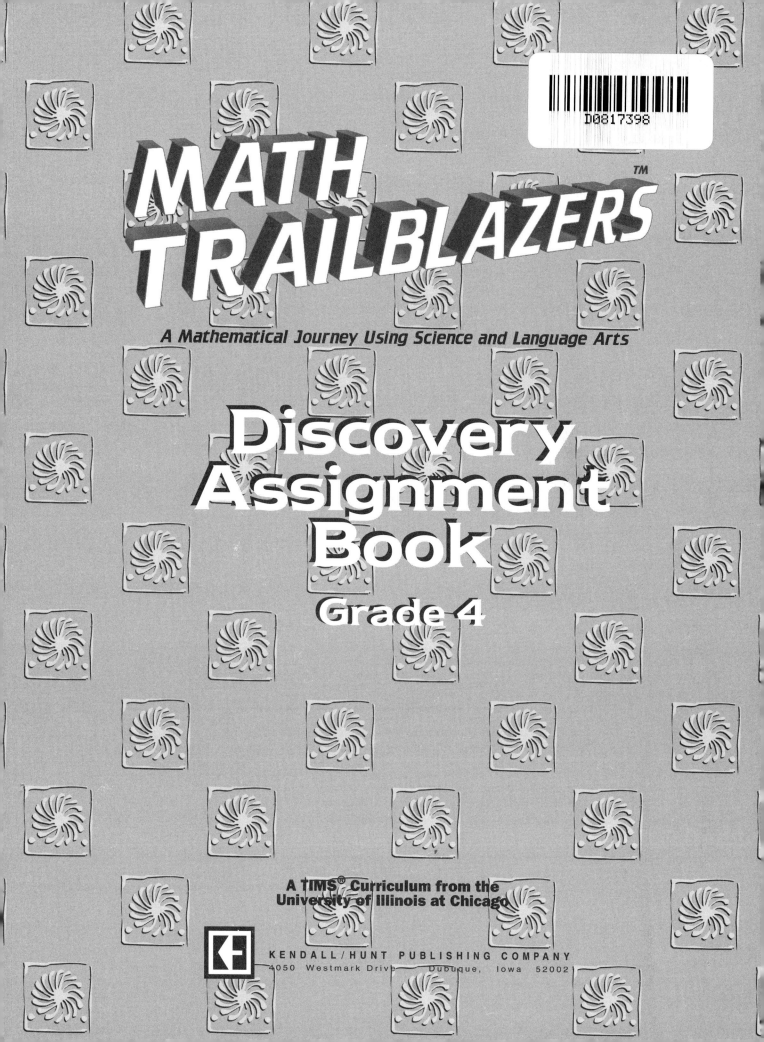

MATH TRAILBLAZERS™

A Mathematical Journey Using Science and Language Arts

Discovery Assignment Book

Grade 4

A TIMS® Curriculum from the
University of Illinois at Chicago

KENDALL/HUNT PUBLISHING COMPANY
4050 Westmark Drive   Dubuque, Iowa 52002

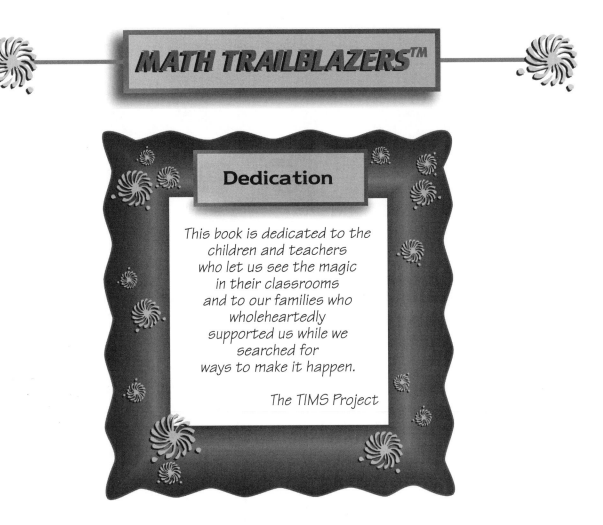

MATH TRAILBLAZERS™

Dedication

This book is dedicated to the
children and teachers
who let us see the magic
in their classrooms
and to our families who
wholeheartedly
supported us while we
searched for
ways to make it happen.

The TIMS Project

UIC The University of Illinois
at Chicago

This material is based on work supported by the National Science Foundation under grant
No. MDR 9050226 and the University of Illinois at Chicago. Any opinions, findings, and
conclusions or recommendations expressed in this publication are those of the authors
and do not necessarily reflect the views of the granting agencies.

Printed in the United States of America
10   9   8   7

Table of Contents

Additional student pages may be found in the *Unit Resource Guide,*
Student Guide, or the *Adventure Book.*

Table of Contents

Additional student pages may be found in the *Unit Resource Guide,*
Student Guide, or the *Adventure Book.*

Dear Parents,

***MATH TRAILBLAZERS*™** is based on the belief that all children deserve a challenging mathematics curriculum and that mathematics is best learned through solving many different kinds of problems. The program provides a careful balance of concepts and skills. Traditional arithmetic skills and procedures are covered through their repeated use in problems and through distributed practice.

***MATH TRAILBLAZERS*™**, however, offers much more. Students using this program will become proficient problem solvers, will know how to approach problems in many different ways, will know when and how to apply the mathematics they have learned, and will be able to communicate clearly their mathematical knowledge. They will learn more mathematics than in a traditional program—computation, measurement, geometry, data collection and analysis, estimation, graphing, patterns and relationships, mental arithmetic, and simple algebraic ideas are all an integral part of the curriculum. They will see connections between the mathematics learned in school and the mathematics used in everyday life. And, they will enjoy and value the work they do in mathematics.

This curriculum was built around national recommendations for improving mathematics instruction in American schools and the research that supported those recommendations. It has been extensively tested with thousands of children in dozens of classrooms over five years of development. ***MATH TRAILBLAZERS*™** reflects our view of a complete and well-balanced mathematics program that will prepare children for a world in the 21st century where proficiency in mathematics will be a necessity. We hope that you enjoy this exciting approach to learning mathematics as you watch your child's mathematical abilities grow throughout the year.

Philip Wagreich

Philip Wagreich
Teaching Integrated Mathematics and Science Project
University of Illinois at Chicago
Chicago, Illinois

UNIT 1

Data About Us

Unit 1: Home Practice

Part 1. Practice

Solve the following addition problems. Try to solve the problems without paper and pencil. Be prepared to share your solution strategies.

1. 8 + 3 + 5 = _____

2. 9 + 7 + 5 = _____

3. 6 + 8 + 6 = _____

4. 4 + 8 + 9 = _____

5. 70 + 30 = _____

6. 60 + 20 + 30 = _____

7. 50 + 70 = _____

8. 30 + 50 + 70 = _____

9. 20 + 85 = _____

10. 10 + 80 + 15 = _____

Part 2. Variables and Values

Look around your house and find four variables. Find two numerical variables and two categorical variables. Then, name some values for each of your variables. Bring this back to class and be prepared to discuss and compare your findings with your classmates. For example: Type of drinks is a categorical variable. Some values for this variable are iced tea, milk, and fruit juice.

1. Variable: _____ numerical or categorical (circle one)

   Values of your variable: _____

2. Variable: _____ numerical or categorical (circle one)

   Values of your variable: _____

3. Variable: _____ numerical or categorical (circle one)

   Values of your variable: _____

4. Variable: _____ numerical or categorical (circle one)

   Values of your variable: _____

Part 3. Finding the Median

1. Mr. Lewis's fourth-grade class did an experiment with colored candies. Five students took a handful of candy. They pulled 12, 6, 5, 3, and 7 pieces of candy. What is the median number of candies pulled? Show how you decided.

2. Lee Yah measured the hand lengths of the people in her family. Her grandmother's hand measured 15 cm in length. Her two sisters' hand lengths were 12 cm and 10 cm. Her mother's hand length was 14 cm. Her father's hand length was 18 cm. Lee Yah's hand measured 12 cm. What is the median hand length in Lee Yah's family? Show how you decided.

3. The fourth grade soccer team at Bessie Coleman School practices after school. This week they practiced for 45 minutes on Monday, 30 minutes on Tuesday, an hour on Thursday, and 40 minutes on Friday. They skipped practice on Wednesday. What is the median number of minutes they practiced for the five days? Show how you decided.

4. Eight of Mrs. Dewey's students stayed after school to help her decorate her bulletin boards. She gave each student a box of raisins as a treat. Each student counted the number of raisins in his or her box. Here is their data: 23, 27, 22, 26, 21, 27, 25, and 23. Based on the students' data, what is the median number of raisins found in a box? Show how you decided.

Part 4. Measuring in Inches

You will need an inch/cm ruler to complete this assignment. Estimate the length in inches of four objects in your home. Then, measure each object to the nearest inch. Complete the data table below.

Object	Estimate (in inches)	Actual (in inches)

Part 5. Inches and Centimeters

1. Which is longer, 1 centimeter or 1 inch? _____ Using a ruler, draw a line that is 1 centimeter long. Draw another line, 1 inch long. Label each line with its measurement.

2. Which is longer, 5 centimeters or 3 inches? _____ Using a ruler, draw a line that is 5 centimeters long. Draw another line, 3 inches long. Label each line with its measurement.

3. **A.** Which is longer, 40 centimeters or 13 inches? _____
   B. How did you decide?

Part 6. The Students in Room 204

Solve each of the following problems. Show how you solved each problem. If you need additional space, use a clean sheet of paper.

1. Ming said, "I collect baseball cards. If I collect 30 more, I'll have 250 cards." How many cards does Ming have now?

2. Keenya likes to listen to music while she practices her tumbling routines for gymnastics class. She listened to two complete cassette tapes while practicing today. There are about 10 songs on each tape, and each song is between 2 and 3 minutes long. About how long did Keenya practice?

3. Irma likes to read. She has two weeks to read the books she took out from the library. One book has 158 pages. The other book has 76 pages.

   A. How many pages are in the two books?
   B. If she reads about ten pages a day, can she finish the two books in two weeks?

4. Maya's family went on a three-day bike trip last week. They biked 36 miles on Friday, 33 on Saturday, and 45 on Sunday. How many miles did they bike in all?

5. It is 4:30 now. Nila's dinner will be ready at 5:15. Nila wants to play her new computer basketball game for twenty minutes. However, she needs fifteen minutes to walk the dog and about seven minutes to set the table. Will Nila be ready for dinner on time? Explain your answer.

6. Write a word problem that describes something about yourself. Write the answer and show how you solved the problem.

Plotting Points Pictures

Number the axes on a sheet of *Centimeter Grid Paper* by ones. Plot the data shown below. Use a ruler to connect the points with lines as you plot each point. When the data table says stop, start at the next new point, but do not connect the previous point.

Picture 1

Horizontal Axis	Vertical Axis	Horizontal Axis	Vertical Axis	Horizontal Axis	Vertical Axis
1	1	14	13	0	10
2	3	12	13	0	5
4	5	10	11	2	9
5	7	11	13	3	10
6	8	15	15	5	9
5	6	10	15	4	8
5	4	9	13	3	6
3	3	8	14	1	4
2	1	8	16	1	1
4	2	10	17	**S T O P**	
6	2	12	16	12	12
7	5	15	17	10	10
8	3	12	18	10	8
7	0	10	19	12	6
10	3	7	18	**S T O P**	
9	6	6	16	13	11
10	4	6	14	12	11
12	2	7	12	12	10
15	1	6	13	13	10
14	3	4	17	13	11
12	4	2	19	**S T O P**	
11	5	0	19	13	8
10	7	3	16	12	8
12	5	5	12	12	7
14	5	6	11	13	7
15	7	4	12	13	8
15	11	2	12	**S T O P**	

Number the axes on a sheet of *Centimeter Grid Paper* by tens. The horizontal axis must go up to 150 and the vertical axis must go up to 200. Plot the data shown below on grid paper. Use a ruler to connect the points with lines as you plot each point.

Picture 2

Horizontal Axis	Vertical Axis
45	0
85	0
125	40
130	80
70	80
70	90
150	90
70	180
105	185
70	200
70	80
60	80
60	200
0	90
60	90
60	80
0	80
5	40
45	0

Arm Span vs. Height

UNIT 2

Geometric Investigations: A Baseline Assessment Unit

Unit 2: Home Practice

Part 1. Practice

1. 60 + 30 = _____

2. 90 + 30 = _____

3. 60 + 60 + 60 = _____

4. 80 – 60 = _____

5. 100 – 70 = _____

6. 130 – 60 = _____

7. 150 – 80 = _____

8. 150 – 90 = _____

9. 180 + 30 = _____

10. 210 – 20 = _____

Part 2. Reading Bar Graphs

Room 202 collected data on the types of pizza they like. Their graph is shown here. Write your answers on a separate sheet of paper.

1. How many students like only cheese on their pizza?

2. How many more students prefer a cheese pizza than a sausage pizza?

3. How many students like pepperoni on their pizza?

4. How many students are in Room 202?

5. Is Type of Pizza a categorical or numerical variable?

6. Is Number of Students a categorical or numerical variable?

7. Which variable did Room 202 graph on the horizontal axis?

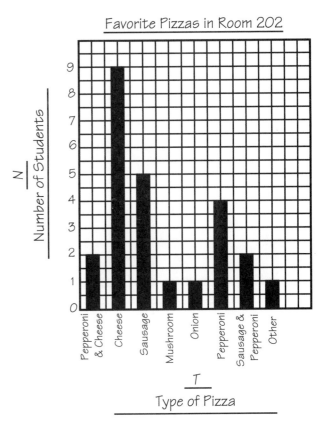

Favorite Pizzas in Room 202

Part 3. Area and Perimeter

1. Find the area and perimeter of the shapes. Include the unit of measurement.

1 sq cm

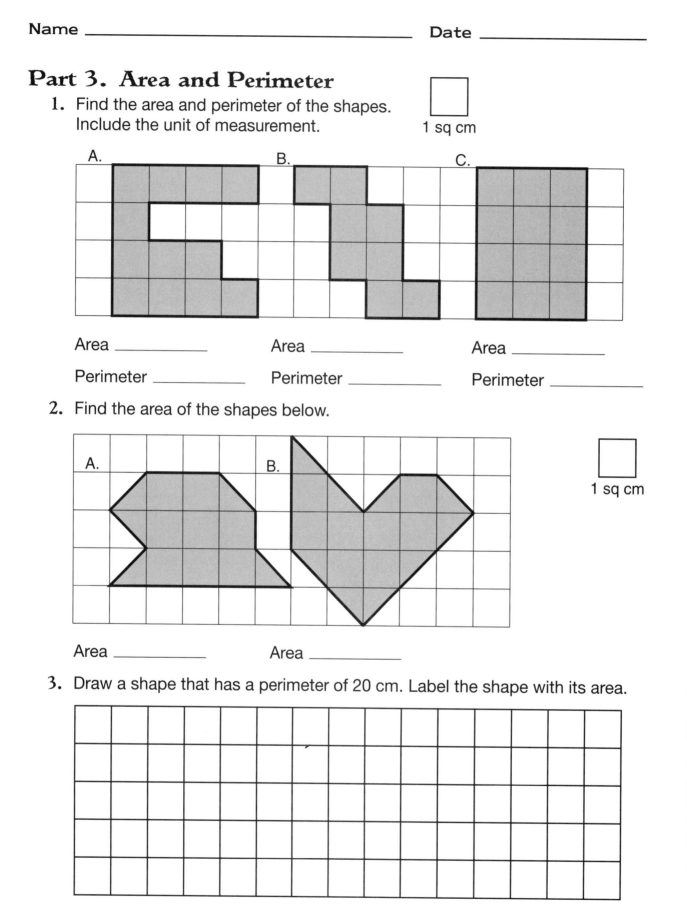

A.

B.

C.

Area _____

Area _____

Area _____

Perimeter _____

Perimeter _____

Perimeter _____

2. Find the area of the shapes below.

1 sq cm

A.

B.

Area _____

Area _____

3. Draw a shape that has a perimeter of 20 cm. Label the shape with its area.

Part 4. Subtraction Practice

Do the following problems in your head.

1. 11 – 9 = _____
2. 13 – 4 = _____
3. 12 – 3 = _____
4. 13 – 8 = _____
5. 12 – 8 = _____

6. 101 – 3 = _____
7. 92 – 4 = _____
8. 42 – 39 = _____
9. 171 – 167 = _____
10. 134 – 5 = _____

Part 5. As Time Passes

1. Draw a right angle.

2. Draw an acute angle.

3. Draw an acute angle that is about 45°.

4. Draw an obtuse angle.

5. What kind of an angle do hands on the clock form at:

A. 9:00? _____

B. 12:05? _____

C. 3:30? _____

D. 12:25? _____

Part 6. Shopping with Grandmother

Solve the following problems. Show how you solved each one using pictures or words.

1. Ming went to the mall with his grandmother. They left the house at 8:00 A.M. They returned home 6 hours later. When did they get home?

2. The bus fare to the mall was 85¢ for his grandmother and 75¢ for Ming. What was the total bus fare to and from the mall? (Remember, they need to come home too!)

3. The mall has two floors. The first floor has 62 stores. The second floor has 48 stores. How many stores are in the mall?

4. The mall newspaper claims that nearly half the stores are participating in a fall sale. About how many stores are participating in the fall sale?

5. Ming and his grandmother shopped in 12 stores. How many stores did they not visit?

6. On the way home, they stopped in the grocery store. Ming went to the deli counter to buy lunch meat. He took a number from the counter which gave his turn in line. The girl behind the counter was waiting on Number 54. Nine more people needed to be served before it was Ming's turn. What number did Ming have?

7. When Ming got home, he and his grandmother played a basketball game on the computer. His grandmother won the game! Her team scored 17 more points than Ming's team. If Ming's team scored 74 points, how many points did his grandmother's team score?

Perimeter-Area Puzzles

Use square-inch tiles to measure the perimeter and area for each shape. Record your measurements on the data table at the bottom of the next page.

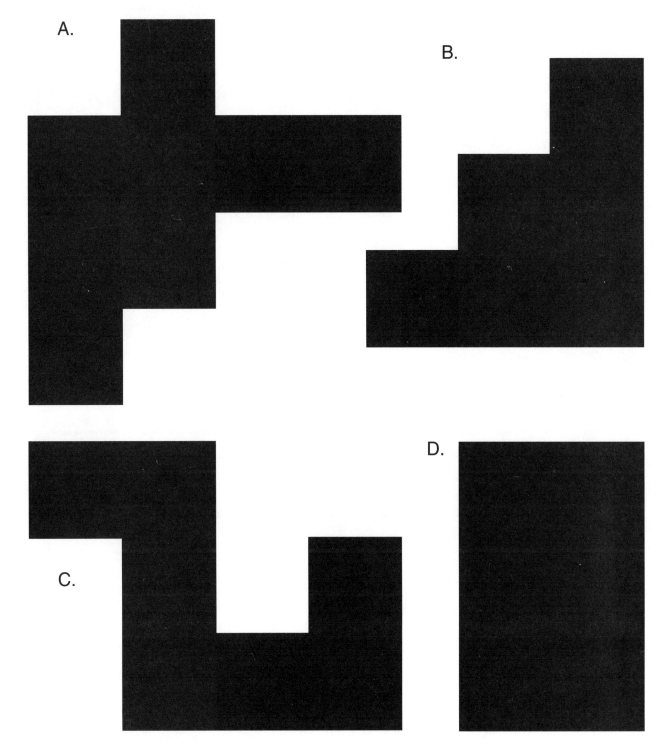

E.

F.

G.

Shape	Perimeter in Inches	Area in Sq Inches
A		
B		
C		
D		
E		
F		
G		

Pattern Block Outlines

Triangle

Square

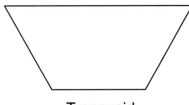

Trapezoid

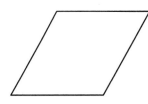

Blue Rhombus

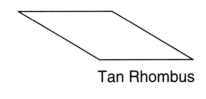

Tan Rhombus

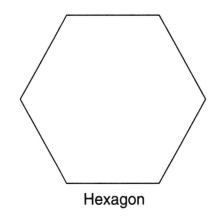

Hexagon

Angle Measures in Pattern Blocks

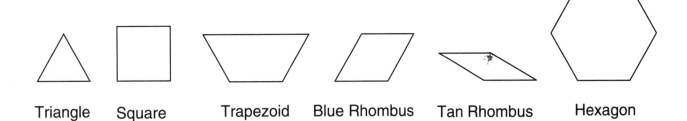

Triangle   Square   Trapezoid   Blue Rhombus   Tan Rhombus   Hexagon

Complete the following table.

Shape	Number of Sides	Number of Angles	Angle Measures
Square			
Triangle			
Trapezoid			
Hexagon			
Blue Rhombus			
Tan Rhombus			

UNIT 3

Numbers and Number Operations

Name _____ Date _____

Unit 3: Home Practice

Part 1. Triangle Flash Cards: 5s and 10s

Study for the quiz on the multiplication facts for the 5s and 10s. Take home your *Triangle Flash Cards: 5s* and *10s* and your list of facts you need to study.

Here's how to use the flash cards. Ask a family member to choose one flash card at a time. He or she should cover the corner containing the highest number. This number will be the answer to a multiplication fact. Multiply the two uncovered numbers.

Study the math facts in small groups. Choose to study 8 to 10 facts each night. Your teacher will tell you when the quiz on the 5s and 10s will be.

Part 2. Addition and Subtraction

Use paper and pencil to solve the following problems.

1.	644	2.	76	3.	386	4.	196
	+ 53		+ 29		− 21		− 77

5.	938	6.	4015	7.	5048	8.	4653
	− 449		+ 488		− 274		+ 5664

Choose one problem. Be ready to explain how you can tell if the answer is reasonable.

Part 3. People and Prices

1. Some of the workers at the TIMS Candy Company went to the fruit stand for lunch. Maggie bought a plum for 29¢ and an apple for 39¢. If she pays with $1.00, how much change should she receive?

2. A comic book at the used book sale costs 5¢. Special edition comic books cost 10¢ each. How much does Shannon need to pay if she wants to purchase 3 regular comic books and 4 special edition comic books?

3. Roberto's father is a salesman for the TIMS Candy Company. At the end of the year, Roberto's father earned a bonus of $565. This was $180 more than last year's bonus.

   A. How much did Roberto's father receive as a bonus last year?

   B. How can you be sure your answer is reasonable?

4. Ana went to the movies. She only had dimes with her. If the movie costs $1.75, how many dimes does she need to give the cashier? Explain.

5. Jessie went with her family to a two-day fall festival. On the first day, 4367 adults attended the fest and 4587 children attended.

   A. The newspaper reported the actual attendance. Calculate the actual attendance.

   B. About how many more people need to attend on the second day to reach a total of 10,000 people at the festival?

Triangle Flash Cards: 10s

- Work with a partner. Each partner cuts out the 9 flash cards and lightly colors the highest number on each card.
- Your partner chooses one card at a time and covers the corner containing the highest number.
- Multiply the two uncovered numbers.
- Divide the used cards into three piles: those that you know and can answer quickly, those that you can figure out, and those that you need to learn.
- Practice the last two piles again. Then, make a list of the facts you need to practice at home.
- Repeat the directions for your partner.

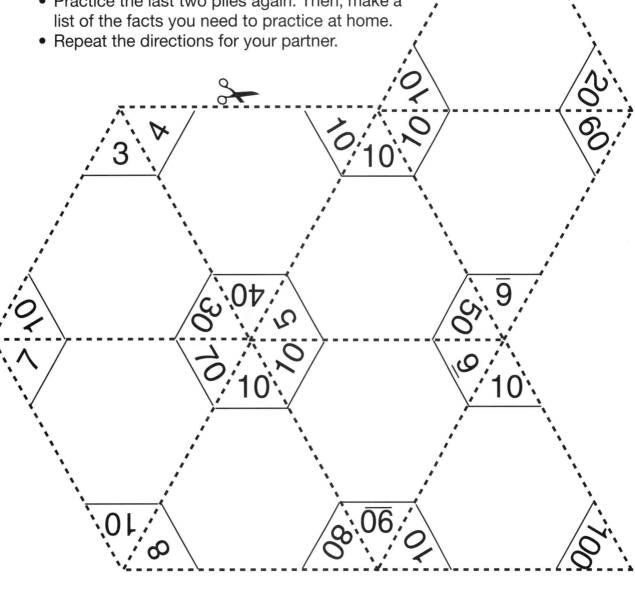

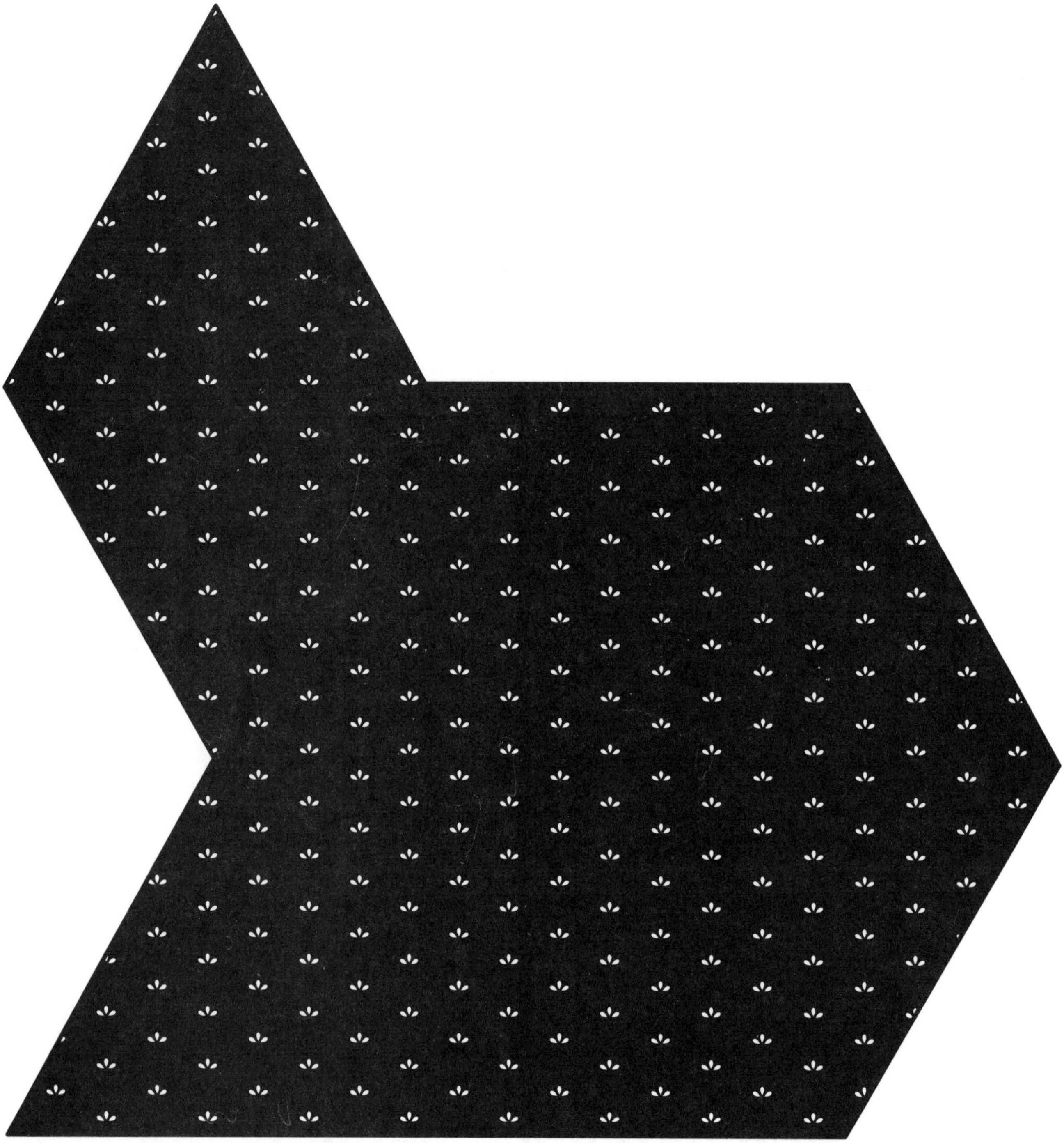

Triangle Flash Cards: 5s

- Work with a partner. Each partner cuts out the 9 flash cards and lightly colors the highest number on each card.
- Your partner chooses one card at a time and covers the corner containing the highest number.
- Multiply the two uncovered numbers.
- Divide the used cards into three piles: those that you know and can answer quickly, those that you can figure out, and those that you need to learn.
- Practice the last two piles again. Then, make a list of the facts you need to practice at home.
- Repeat the directions for your partner.

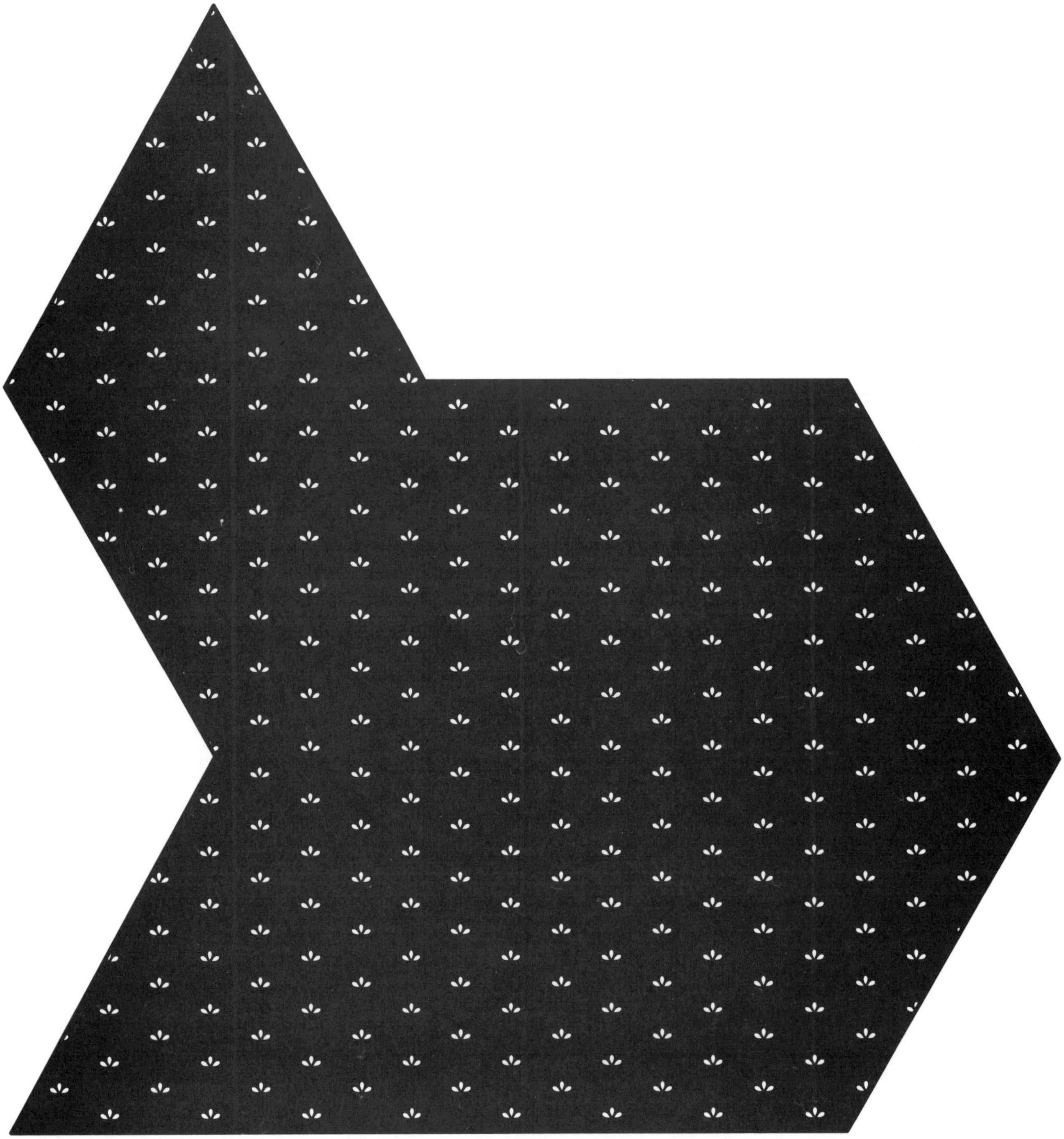

Number Line

Cut out the number line pieces below and tape or glue them at the tabs.

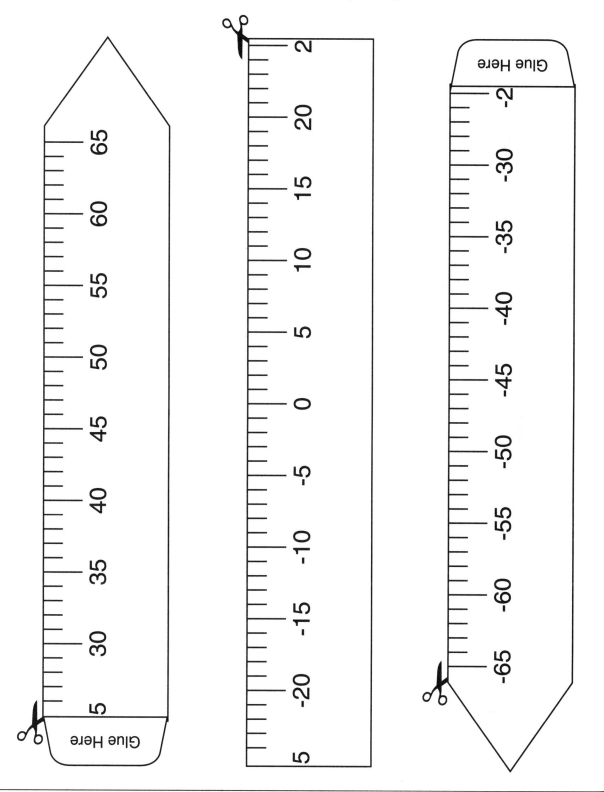

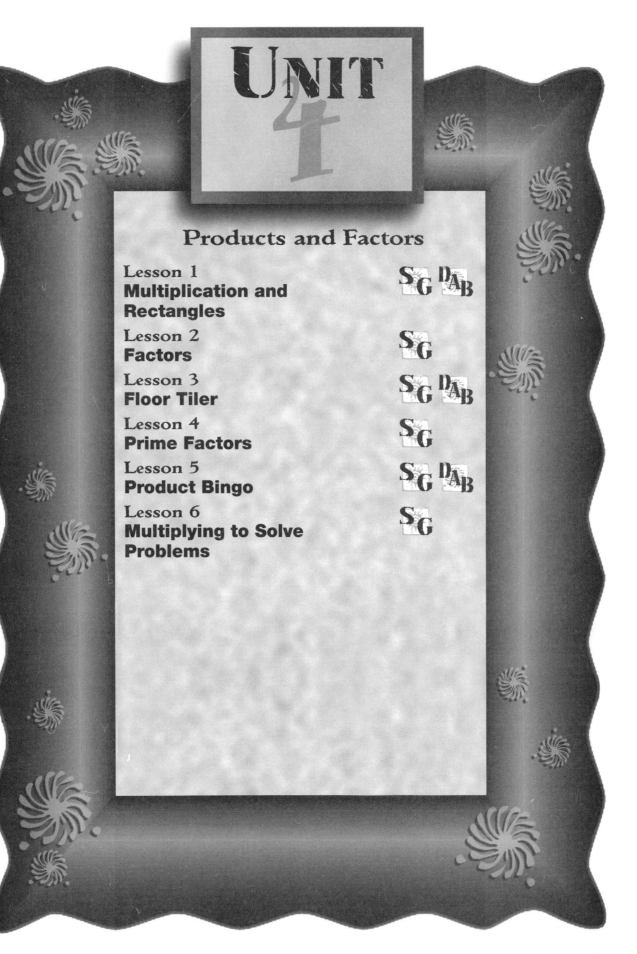

UNIT 4

Products and Factors

Name _____ Date _____

Unit 4: Home Practice

Part 1. Triangle Flash Cards: 2s and 3s

Study for the quiz on the multiplication facts for the 2s and 3s. Take home your *Triangle Flash Cards: 2s and 3s* and your list of facts you need to study.

Here's how to use the flash cards. Ask a family member to choose one flash card at a time. He or she should cover the corner containing the highest number. This number will be the answer to a multiplication fact. Multiply the two uncovered numbers.

Study the math facts in small groups. Choose to study 8 to 10 facts each night. Your teacher will tell you when the quiz on the 2s and 3s will be.

Part 2. Multiplication: Factors, Multiples, Primes, and Squares

To solve the following problems, you may use your *Student Guide* as a reference. See Unit 4 Lessons 1, 2, and 4.

1. Is 34 a multiple of 2? Explain why or why not.

2. Is 3 a factor of 35? Explain why or why not.

3. Name 10 numbers which are multiples of 2.

4. Name 10 numbers which have 3 as a factor.

5. Is 7 a prime number? Why or why not?

6. **A.** $5^2 =$ **B.** $10^2 =$

    C. $2^2 =$ **D.** $3^2 =$

Part 3. Working at the Grocery Store

Choose an appropriate tool to help you solve each of the problems. Use a picture, paper and pencil, or a calculator. Show how you solved each problem.

1. Keenya's sister, Shenika, works at a grocery store. Today she is stocking shelves. She stacks soup cans three cans high. If she makes 6 stacks, how many soup cans will she shelve?

2. Shenika brings in some shopping carts from the parking lot. She makes 4 rows of carts. She tries to place the same number of carts in each row. If she brings in 17 carts, how many carts can she place in each row?

3. Shenika gets paid six dollars an hour. Last week she worked 15 hours. How much did she earn?

4. When Shenika works the cash register on a Saturday, she works nonstop. In the express line, customers may purchase only 10 items or less. When working the express line, she can ring up about 5 customers in 15 minutes. If she works a 6-hour day, about how many customers can she serve?

5. The grocery bill for one of Shenika's customers is $15.52. The customer gives Shenika $20.02. How much change should the customer receive?

6. Grapes are on sale for 69¢ a pound. How much do 3 pounds of grapes cost?

7. For every $3 a customer spends at the grocery store, he or she gets a stamp that can be used for purchasing dishes. One customer bought groceries for herself and an elderly neighbor. The two separate bills were $43 and $28. How many stamps should this customer receive for herself and her neighbor?

Triangle Flash Cards: 2s

- Work with a partner. Each partner cuts out the 9 flash cards and lightly colors the highest number on each card.
- Your partner chooses one card at a time and covers the corner containing the highest number.
- Multiply the two uncovered numbers.
- Divide the used cards into three piles: those that you know and can answer quickly, those that you can figure out, and those that you need to learn.
- Practice the last two piles again. Then, make a list of the facts you need to practice at home.
- Repeat the directions for your partner.

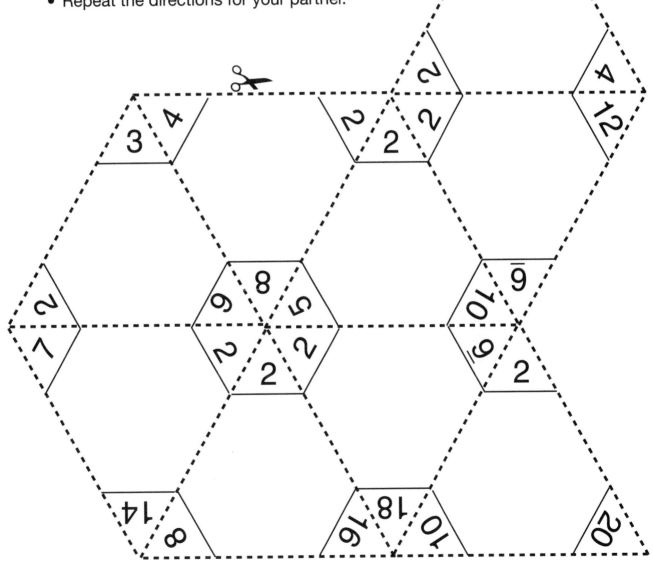

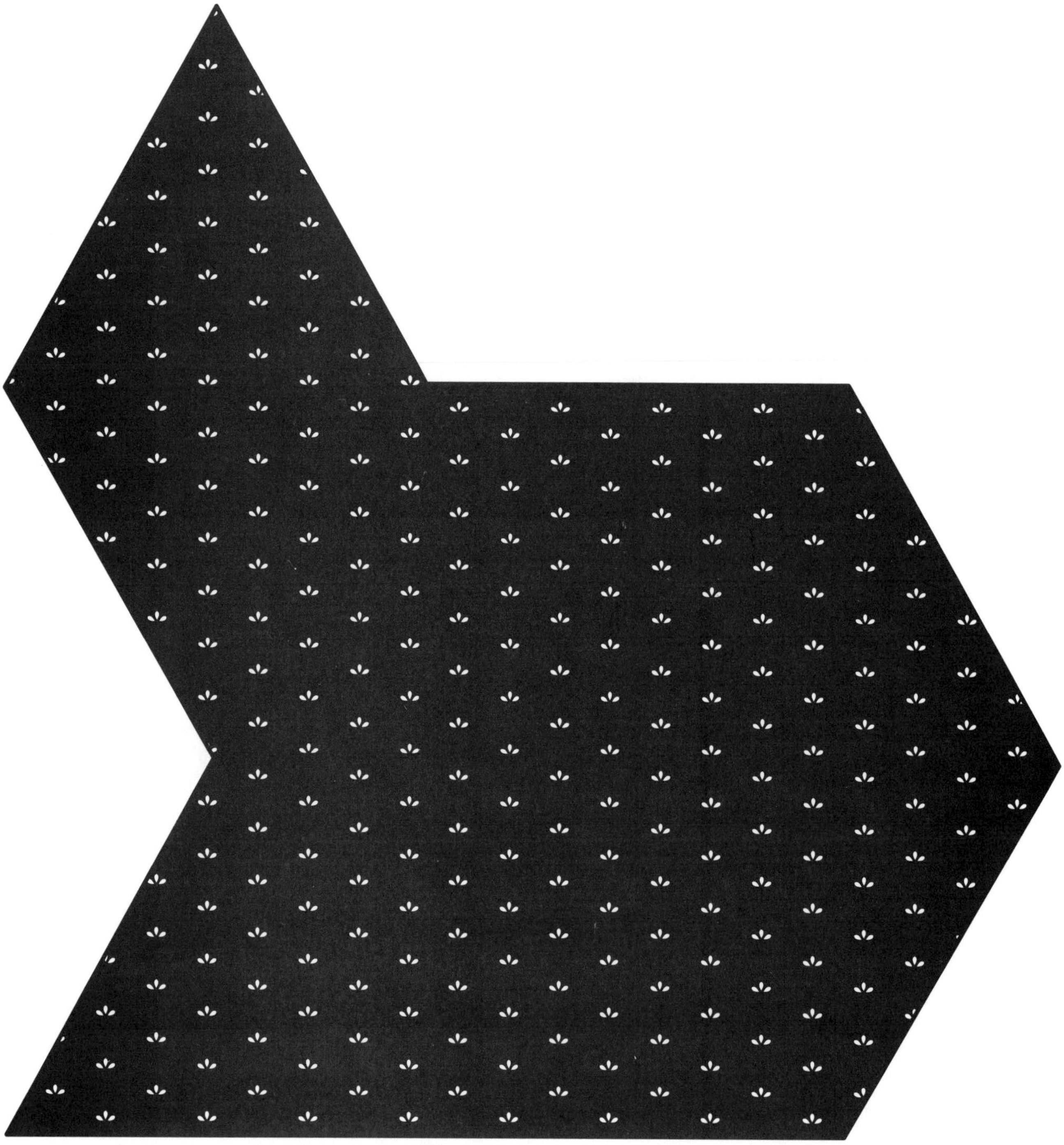

Name _____   Date _____

Triangle Flash Cards: 3s

- Work with a partner. Each partner cuts out the 9 flash cards and lightly colors the highest number on each card.
- Your partner chooses one card at a time and covers the corner containing the highest number.
- Multiply the two uncovered numbers.
- Divide the used cards into three piles: those that you know and can answer quickly, those that you can figure out, and those that you need to learn.
- Practice the last two piles again. Then, make a list of the facts you need to practice at home.
- Repeat the directions for your partner.

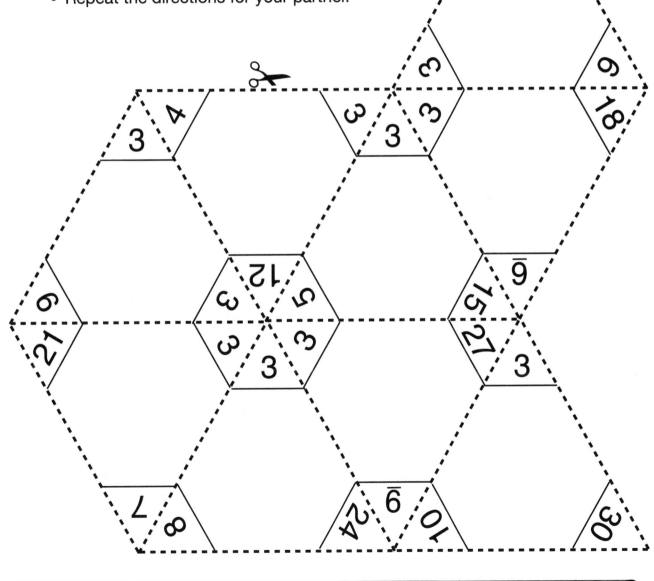

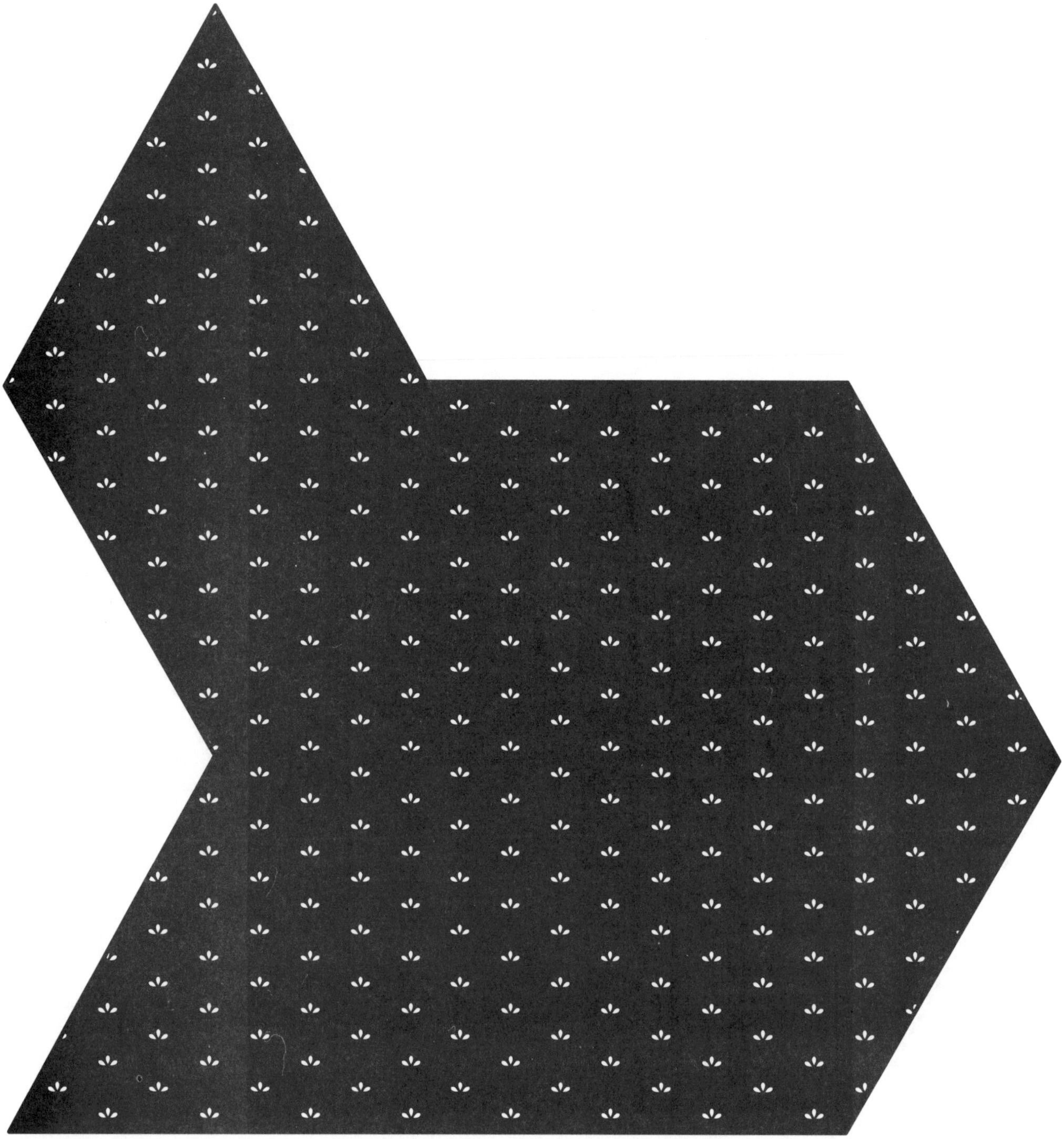

Name _____ Date _____

Rectangles

1	
2	
3	
4	
5	
6	
7	
8	
9	
10	
11	
12	
13	

14	
15	
16	
17	
18	
19	
20	
21	
22	
23	
24	
25	

Spinners 1–4 and 1–10

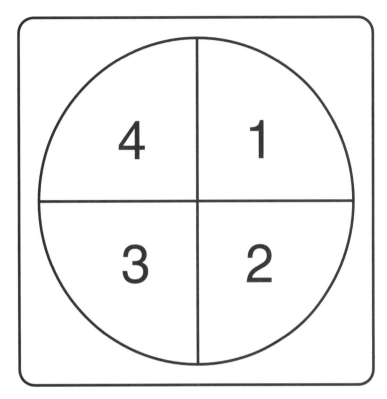

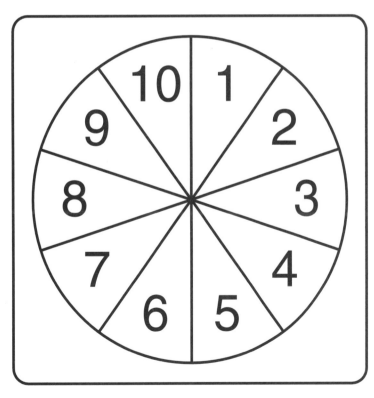

Product Bingo Game Boards

Board 1

40	72	10	35
42	28	20	27
15	45	P	6
30	48	14	56

Board 3

8	24	27	54
20	12	21	32
36	P	14	45
63	18	72	16

Board 2

9	P	22	81
64	13	25	32
15	14	56	29
7	10	4	49

Board 4

4	45	25	81
49	56	6	32
9	64	P	10
15	8	42	48

UNIT 5

Using Data to Predict

Unit 5: Home Practice

Part 1. Triangle Flash Cards: Square Numbers

Study for the quiz on the multiplication facts for the square numbers. Take home your *Triangle Flash Cards: Square Numbers* and your list of facts you need to study.

Here's how to use the flash cards. Ask a family member to choose one flash card at a time. Your helper should cover the corner containing the highest number. This number will be the answer to a multiplication fact. Multiply the two uncovered numbers.

Your teacher will tell you when the quiz on the square numbers will be.

Part 2. Time and Roman Numerals

A table of Roman numerals is provided in Unit 3, Lesson 2 of your *Student Guide.* You may use it as a reference.

1. Skip count by 3s from 3 to 30 using Roman numerals.

III	VI		XII			XXI			
3	6	9		15					

2. Skip count by 20s from 20 to 200 using Roman numerals.

XX	XL				CXX				
20	40					140			

3. What time does each clock show?

A.   B.   C.

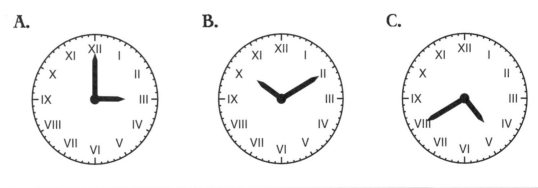

Part 3. Factor Trees and Exponents

Write each of the following numbers as a product of prime numbers. If you need more room to show your work, use a separate sheet of paper.

1. 52   2. 85   3. 224

4. Write each of the following using exponents. Then, find each product.

A. $4 \times 4 \times 2$ **B.** $5 \times 2 \times 5$ **C.** $2 \times 3 \times 2 \times 2$

Part 4. What's Missing?

The letter _n_ stands for a missing number. What number must _n_ be in each number sentence to make the sentence true?

1. $750 + 150 = n$ 2. $839 + 102 = n$ 3. $1034 - 40 = n$

4. $2 + n = 100$ 5. $16 - n = 8$ 6. $n + 21 = 42$

7. $n - 25 = 50$ 8. $11 + n = 24$ 9. $93 - n = 23$

10. $70 - n = 40$ 11. $71 - n = 40$ 12. $15 - n = 9$

Part 5. Making Predictions from a Graph

The data below shows the winning times women swam the 200-meter backstroke at the Olympics. The winning times are given to the nearest second.

Time (in years)	Time (in seconds)
1968	145
1972	139
1976	133
1980	132
1984	132
1988	129
1992	127

1. Plot the data on a piece of *Centimeter Graph Paper.* Scale the vertical axis to 150 seconds. Scale the horizontal axis by fours, from 1960 to 2004. Then, answer Questions 2–7. Use a separate sheet of paper if you need to.

2. **A.** What variable did you plot on the horizontal axis on the graph?

   _____

   B. Is this variable numerical or categorical? _____

3. What variable did you plot on the vertical axis? _____

4. About how many minutes was the winning time for the 200-meter

   backstroke in 1992? _____

5. Do the points lie close to a line? If they do, use a ruler to draw a best-fit line through the points.

6. Describe the graph. What does it tell you about the history of the 200-meter backstroke?

7. Can you use this graph to predict the winning time for the women's 200-meter backstroke in the year 2000? If so, how? What is your prediction?

Part 6. Bouncing Balls

Grace and her lab partner Michael experimented with 3 kinds of balls to find out which one bounced highest. They dropped each type of ball from the same height. Here is their data.

T Type of Ball	*H* Bounce Height (in cm)			
	Trial 1	**Trial 2**	**Trial 3**	**Median**
Basketball	43	41	45	
Kickball	69	65	67	
Tennis Ball	52	51	51	

1. Find the median bounce height for each type of ball. Complete the table with your answers.

2. What is the manipulated variable? Is it a categorical or numerical variable?

3. What is the responding variable? Is it a categorical or numerical variable?

4. Think about these questions before you graph the median bounce height for each type of ball.

    *   What variable will you put on the horizontal axis?
    *   What variable will you put on the vertical axis?
    *   How will you scale and label the axes?
    *   What type of graph is appropriate? A point graph or a bar graph?

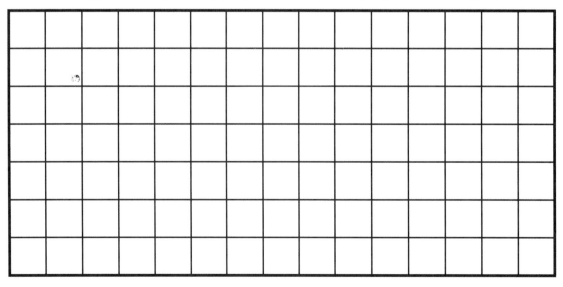

Triangle Flash Cards: Square Numbers

- Work with a partner. Each partner cuts out the 9 flash cards and lightly colors the highest number on each card.
- Your partner chooses one card at a time and covers the corner containing the highest number.
- Multiply the two uncovered numbers.
- Divide the used cards into three piles: those that you know and can answer quickly, those that you can figure out, and those that you need to learn.
- Practice the last two piles again. Then, make a list of the facts you need to practice at home.
- Repeat the directions for your partner.

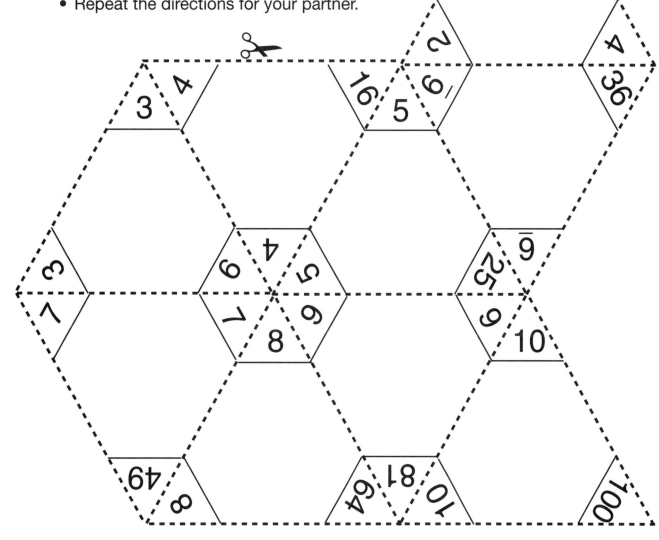

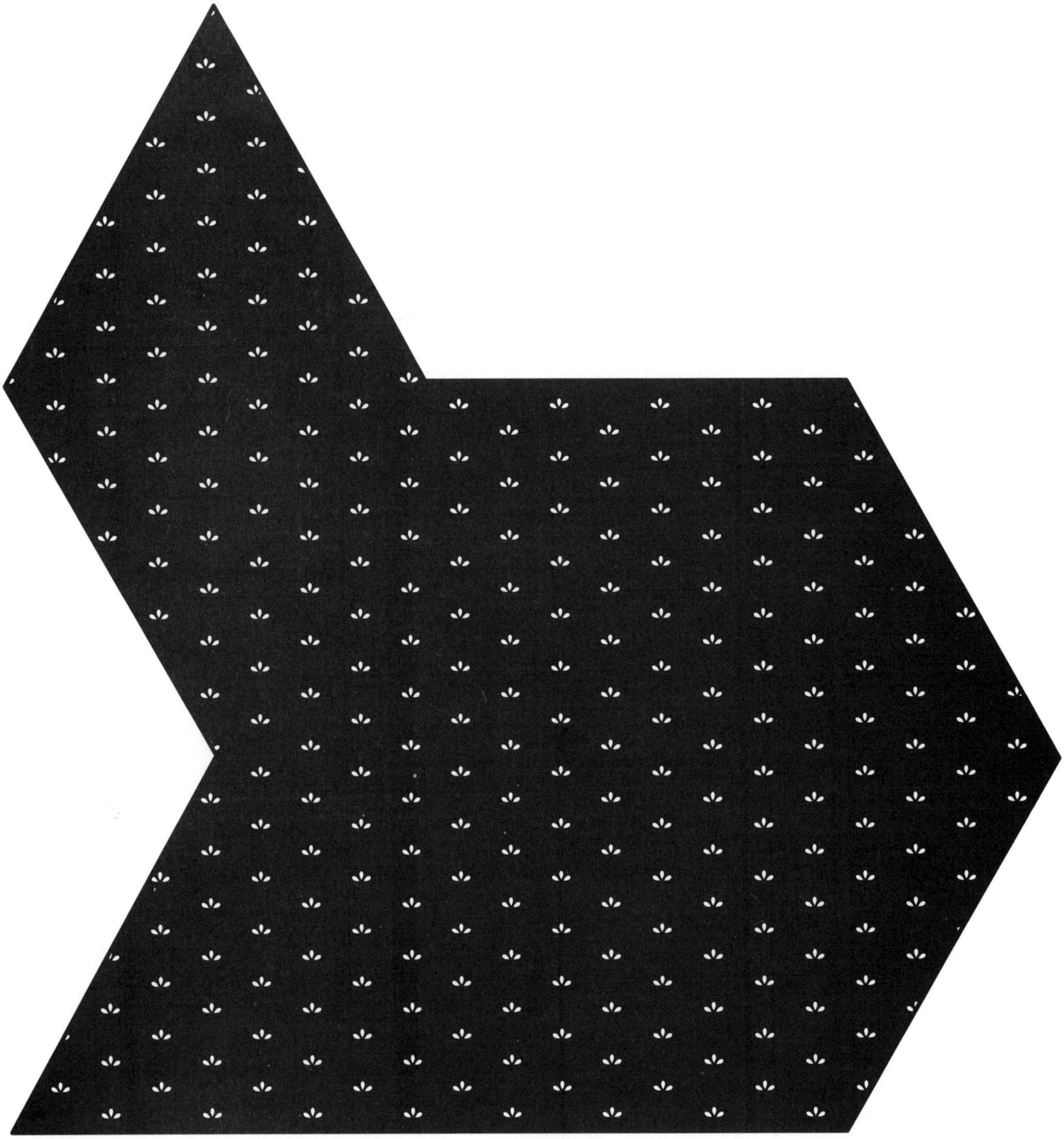

Using Best-Fit Lines

1. Each year, Mrs. Welch, a gym teacher at Bessie Coleman School, records the number of sit-ups each student can do. Nila used her data to make a graph which shows the number of sit-ups she could do each year.

   A. Describe the graph.

   B. If you read the graph from left to right, do the points go uphill or downhill?

   C. What does the graph tell you about the number of sit-ups Nila can do?

   D. Do the points lie close to a straight line? If so, use a ruler to draw a best-fit line.

   E. If possible, predict the number of sit-ups Nila will be able to do when she is 12.

   F. Does knowing Nila's age help you predict the number of sit-ups she can do?

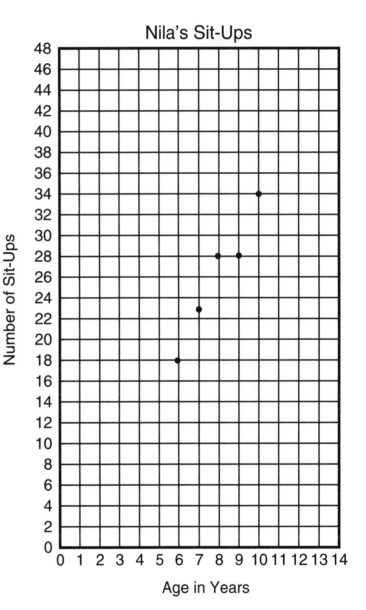

Nila's Sit-Ups

2. Mrs. Welch also records each student's best times for running a mile. John graphed his best times.

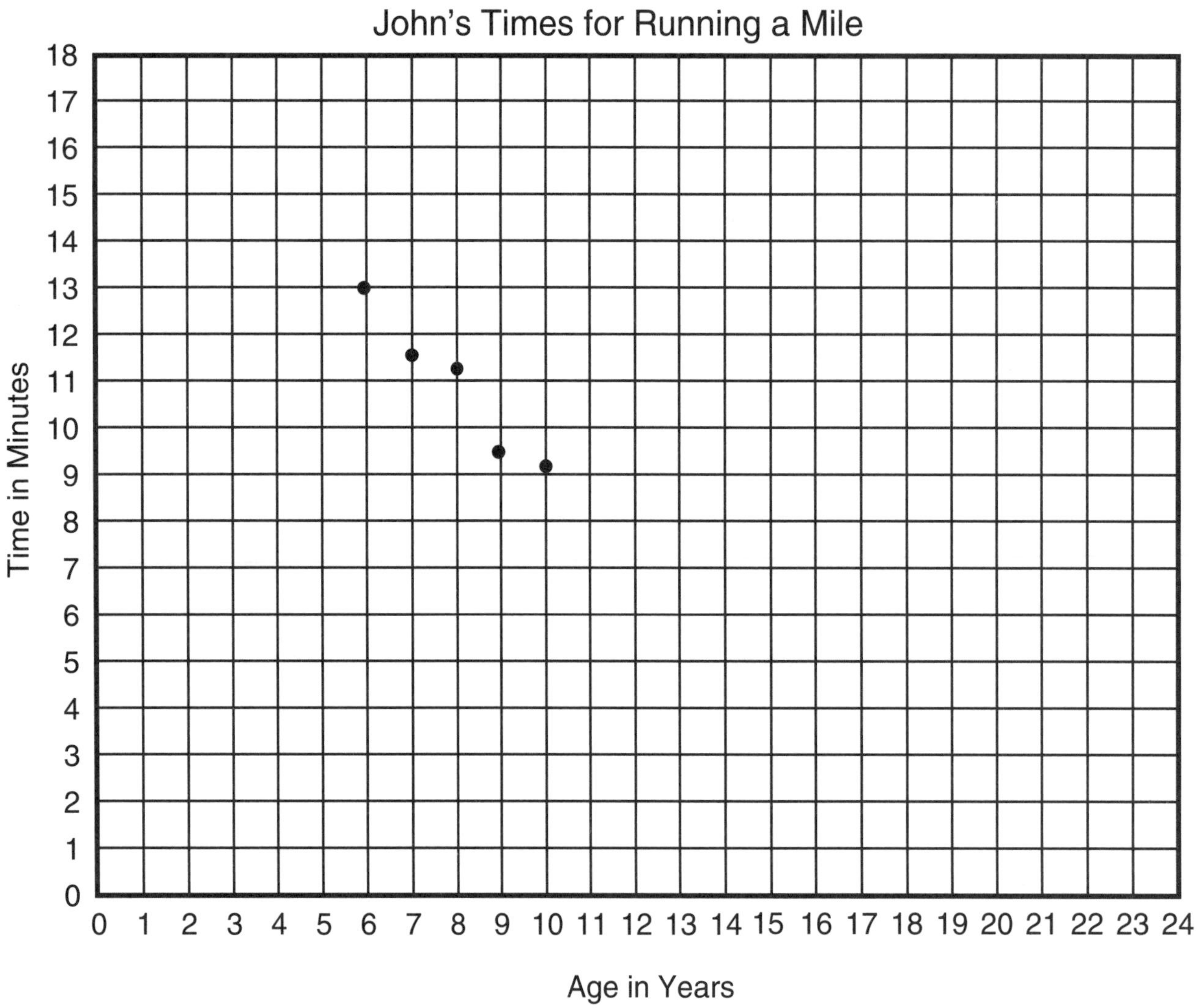

John's Times for Running a Mile

Time in Minutes

Age in Years

A. Describe the graph.

B. Do the points tend to go uphill or downhill?

C. Do the points lie close to a straight line? If so, use a ruler to draw a best-fit line.

D. If possible, predict how long it will take John to run a mile when he is 12.

E. If possible, predict how long it will take John to run a mile when he is 18.

F. Does knowing John's age help you predict his time for running the mile?

3. A fourth-grade class recorded the month each student was born and the number of letters in each student's name. Using the data, the class made the following graph.

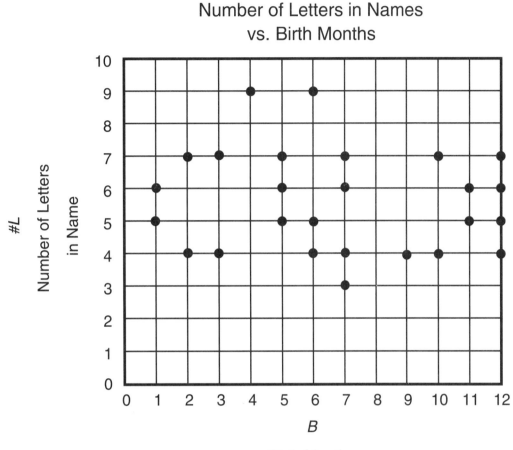

Number of Letters in Names
vs. Birth Months

#L
Number of Letters
in Name

B

Birth Months

A. Describe the graph.

B. Do the points lie close to a straight line? If so, use a ruler to draw a best-fit line.

C. Does knowing the month a student was born help you predict the number of letters in his or her name?

D. If possible, predict the number of letters in a student's name if he or she was born in August (the eighth month).

4. A cookie company wants all the cookies from the factory to be the same. Here is a graph made by a cookie inspector.

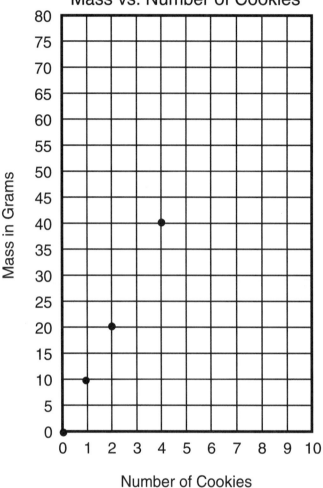

Mass vs. Number of Cookies

Mass in Grams

Number of Cookies

A. Describe the graph.

B. Do the points lie close to a straight line? If so, use a ruler to draw a best-fit line.

C. If possible, predict the mass of 3 cookies.

D. If possible, predict the mass of 5 cookies.

E. Did you use interpolation or extrapolation to answer Question 4C?

5. Doctors measure the head circumference of babies to track their growth.

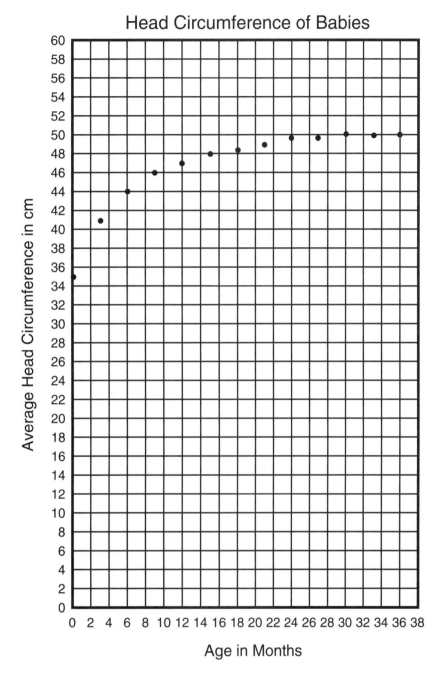

Head Circumference of Babies

A. Describe the graph.

B. If the points lie close to a line, use a ruler to draw a best-fit line.

C. If possible, predict the head circumference of a baby who is four years old.

6. The winning times for the women's Olympic breaststroke swimming competition are shown in this graph.

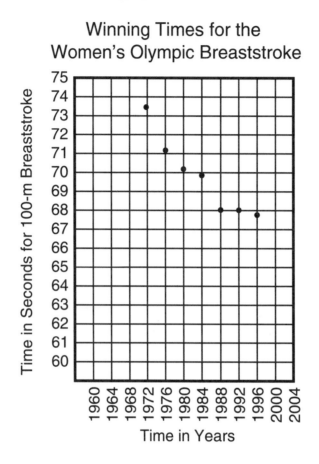

Winning Times for the
Women's Olympic Breaststroke

A. Describe the graph.

B. If the points lie close to a line, use a ruler to draw a best-fit line.

7. Look back at the graphs in Questions 1–6. Which graph gives the most accurate predictions? Explain your choice.

UNIT 6

Place Value Patterns

Unit 6: Home Practice

Part 1. Triangle Flash Cards: 9s

Study for the quiz on the multiplication facts for the nines. Take home your *Triangle Flash Cards: 9s* and your list of facts you need to study.

Here's how to use the flash cards. Ask a family member to choose one flash card at a time. He or she should cover the corner containing the highest number. This number will be the answer to a multiplication fact. Multiply the two uncovered numbers.

Your teacher will tell you when the quiz on the 9s will be.

Part 2. Mixed-Up Multiplication Tables

1. Complete the table. Then, describe any patterns you see.

×	2	3	5	9	10
4					
6		18			
7					
8					

2. *n* stands for a missing number. Find the missing number in each number sentence.

   A. $n \times 7 = 14$ **B.** $3 \times n = 24$ **C.** $n \times 4 = 16$ **D.** $n \times 8 = 80$

   E. $9 \times n = 63$ **F.** $n \times 8 = 64$ **G.** $4 \times n = 36$ **H.** $n \times 5 = 30$

Part 3. Addition and Subtraction Practice

Use paper and pencil to solve the following problems. Use estimation to help you decide if your answers make sense.

A. 4506
  + 8753

B. 5388
  + 9078

C. 9054
  − 2408

D. 7617
  − 4543

E. 3940
  + 6963

F. 10,415
  − 7593

Part 4. Using Estimation

The following table lists the number of people that immigrated to the United States from various countries in 1993. Use the information in the table to estimate the answers to the questions below. Use a separate sheet of paper to show what convenient numbers you chose to work with.

Country	Number of Immigrants	Country	Number of Immigrants
Canada	17,156	India	40,121
China	65,578	Mexico	126,561
Dominican Republic	45,420	Philippines	63,457
El Salvador	26,818	Russia	58,571
Great Britain	18,783	Vietnam	59,614

1. Most immigrants came from which five countries listed in the table? List the countries and their population. List the populations in order from largest to smallest.

2. About how many more people immigrated from Mexico than from China?

3. About how many people immigrated from Canada and Mexico combined?

4. The number of immigrants from Vietnam is about the same as the number from which other country?

5. About how many more people came from Russia than India?

6. In 1993, the number of immigrants from all countries totaled 904,292 people. About how many immigrants are reported in the table above? (*Hint:* Use a calculator to help you with your estimation.)

Part 5. Numbers in the News

Keenya and Nicholas found the following big numbers in the newspaper. Write the numbers in order from smallest to largest in the following place value chart.

4,130,243   7,931,435   39,905   793,027

4,613,378   9835   42,319

Millions			Thousands			Ones		

Write the smallest number in words.

Write the largest number in words.

Part 6. Area and Perimeter: Within 10%

1. Jessie's mother is tiling her kitchen floor. She estimated that her kitchen floor has an area of 95 square feet. She measures the room and finds that the area is 104 square feet.

   A. Is her estimate within 10% of the measured area? How do you know?

   B. Is her estimate of 95 square feet a good one for ordering tile? Why or why not?

2. Frank's father is putting a fence around his backyard. He estimated that the perimeter of his backyard is about 340 feet. He measures the perimeter and finds that it is 311 feet.

   A. Is his estimate within 10% of his measurement? How do you know?

   B. Is his estimate of 340 feet a good one for making a fence? Why or why not?

Part 7. Convenient Numbers

Estimate where each of the numbers (A–D) is located on the following number line. Make a mark on the number line to show each number. Label each mark with the correct letter A, B, C, or D. Then, use the number line to round each number to the nearest ten thousand and nearest hundred thousand.

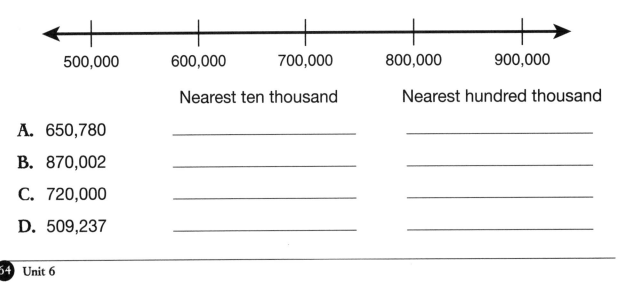

	Nearest ten thousand	Nearest hundred thousand
A. 650,780	_____	_____
B. 870,002	_____	_____
C. 720,000	_____	_____
D. 509,237	_____	_____

Triangle Flash Cards: Nines

- Work with a partner. Each partner cuts out the 9 flash cards and lightly colors the highest number on each card.
- Your partner chooses one card at a time and covers the corner containing the highest number.
- Multiply the two uncovered numbers.
- Divide the used cards into three piles: those that you know and can answer quickly, those that you can figure out, and those that you need to learn.
- Practice the last two piles again. Then, make a list of the facts you need to practice at home.
- Repeat the directions for your partner.

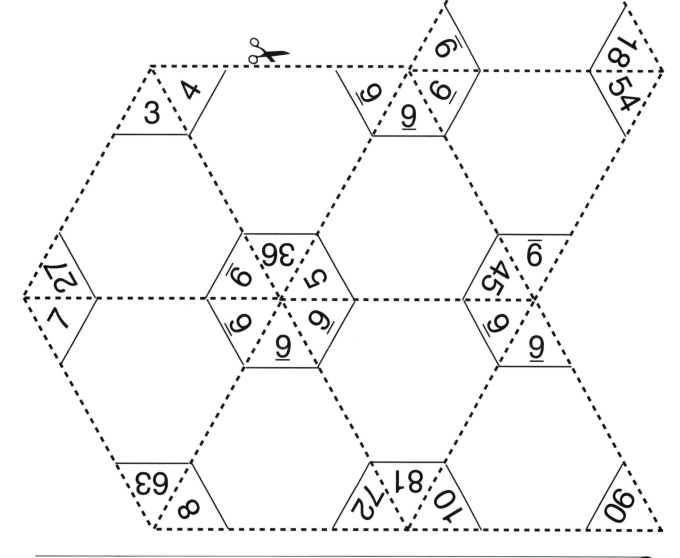

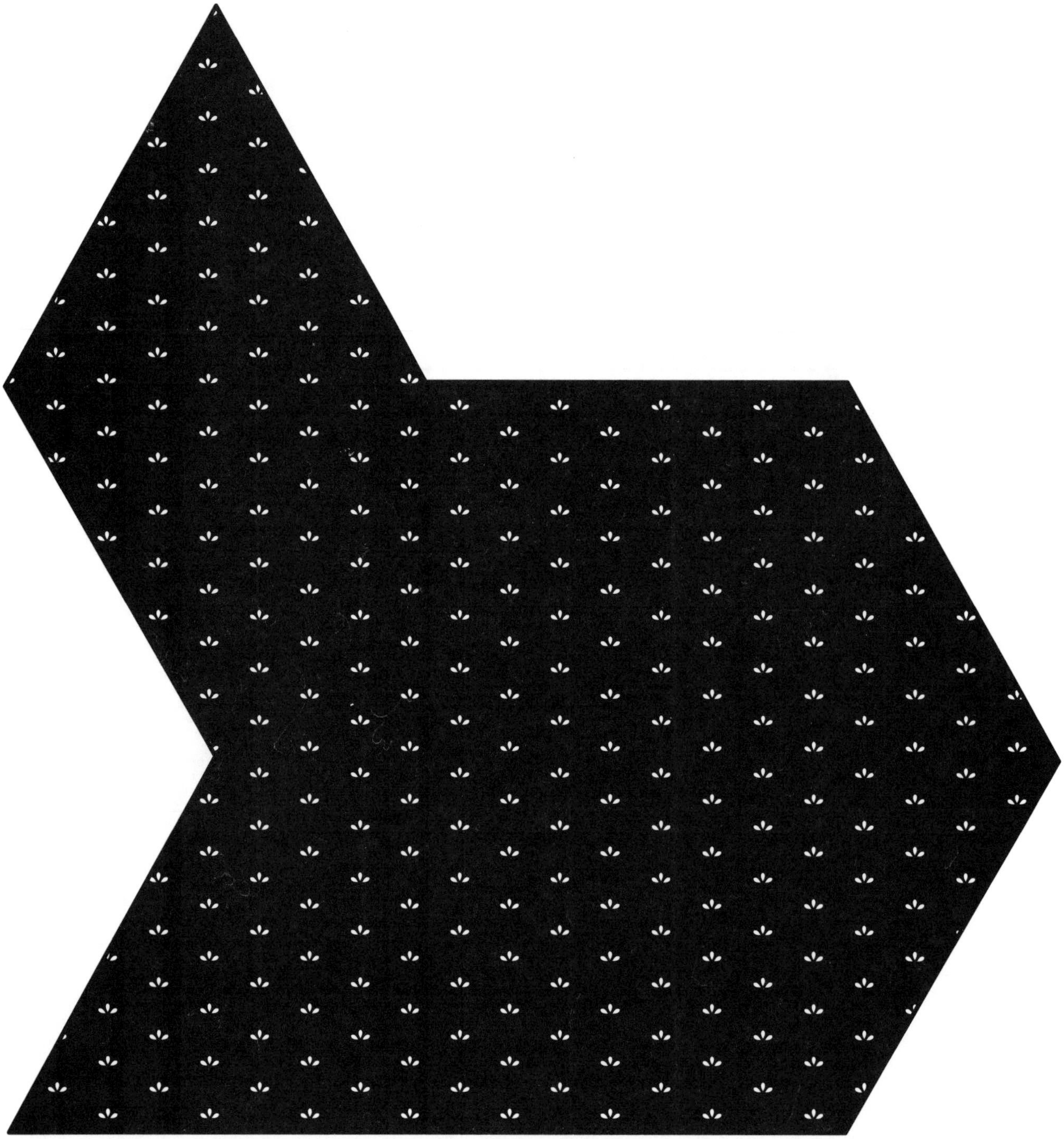

Name _____ Date _____

Place Value Chart

Millions			Thousands			Ones		

Draw, Place, and Read

This game can be played with 2 to 30 players. One player is the "caller." You will need 10 cards numbered 0–9. If you have a large group, the game can be played with teams. Follow these directions:

1. Choose one person to be the caller. For each round, the caller will draw seven numbers. Decide ahead of time if you will replace each number after it has been recorded or if you will use each number only once in each round.

2. After each draw, players record the digit that was drawn in any column on the first row of the place value chart below. Once a digit has been recorded, it cannot be moved.

3. After all seven draws, the person who makes and reads the highest number earns a point. Move to the next row on the place value chart for the next round.

As an extra challenge, you can agree to use 8 or 9 draws instead of 7 for each round.

Millions' Period			Thousands' Period			Ones' Period		

10% Chart

Name on Jar	Object	N Actual Number	N ÷ 10	10% of the Number	Range

9 to 5 War Cards

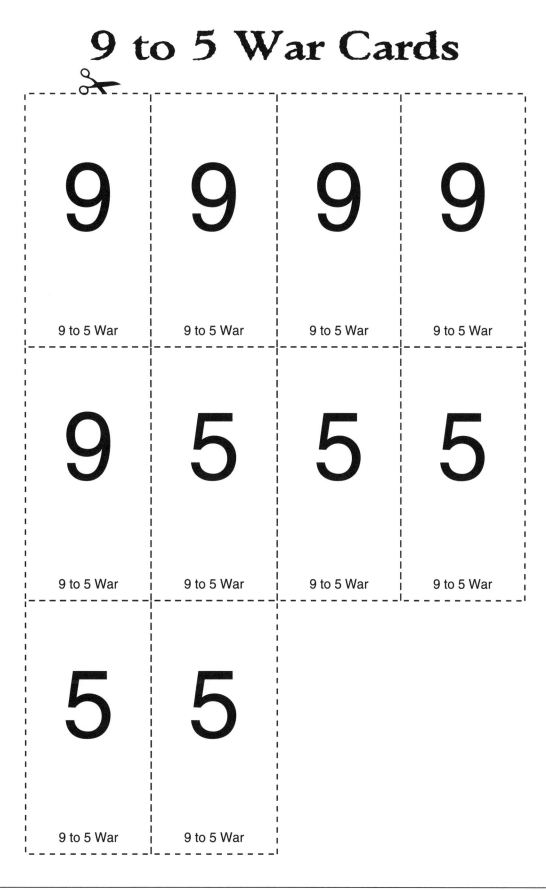

9 to 5 War Cards

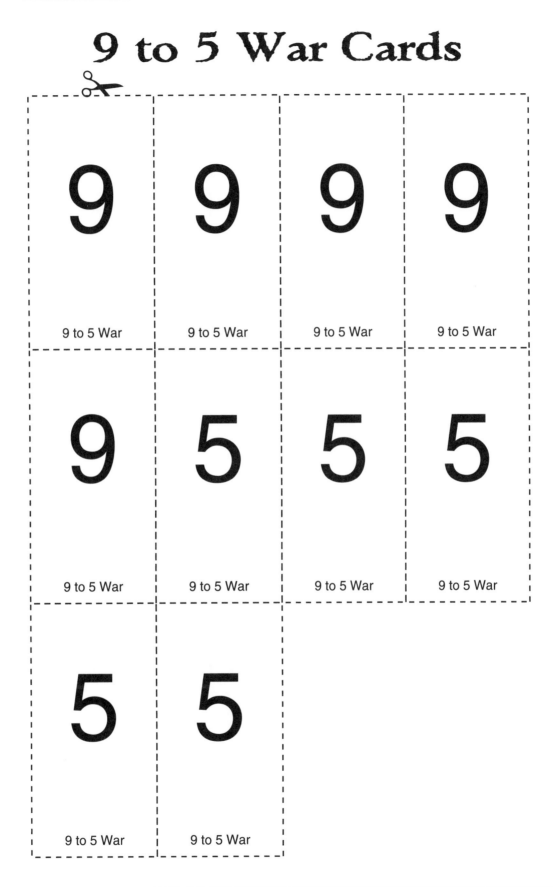

UNIT 7

Patterns in Multiplication

Unit 7: Home Practice

Part 1. Triangle Flash Cards: The Last Six Facts

Study for the quiz on the multiplication facts for the last six facts. Take home your *Triangle Flash Cards: The Last Six Facts* and your list of facts you need to study.

Here's how to use the flash cards. Ask a family member to choose one flash card at a time. Your partner should cover the corner containing the highest number. This number will be the answer to a multiplication fact. Multiply the two uncovered numbers.

Your teacher will tell you when the quiz on the last six facts will be.

Part 2. Order of Operations

1. Remember the proper order of operations as you do the following problems. You may use a calculator, but be sure you follow the correct order of operations even if your calculator does not.

   A. $7 \times 2 + 5 =$ _____   **B.** $8 + 4 \times 3 =$ _____

   C. $7 + 24 \div 3 =$ _____   **D.** $7 \times 4 + 5 \times 2 =$ _____

   E. $8 \times 6 - 3 \times 3 =$ _____   **F.** $36 \div 9 + 6 \times 7 =$ _____

   G. $7 + 9 \times 8 - 5 =$ _____   **H.** $100 - 49 \div 7 + 10 =$ _____

2. Play *Operation Target*. Use the numbers 1, 2, 3, and 4 and the four operations to make as many different whole numbers as you can. You need paper, a pencil, and a calculator. In each number sentence, you must use each of the four digits exactly once. You can use operations more than once or not at all. For example: To make 10 you could write: $4 \times 1 + 2 \times 3 = 10$. Use a separate sheet of paper to write your number sentences for each of the numbers you make.

Part 3. Division

Use the numbers listed below to answer the following questions.

567   85,680   289   27,786   1028   10,782

1. Which numbers are divisible by 2? How did you decide?

2. Which numbers are divisible by 3? How did you decide?

3. Which are divisible by 6? How did you decide?

4. Which are divisible by 5 and 10? How did you decide?

5. Which are divisible by 9? How did you decide?

Part 4. Addition and Subtraction Practice

Solve the following problems in your head or using paper and pencil.

A. 210 + 42 = _____   B. 360 + 18 = _____

C. 350 + 35 = _____   D. 480 + 36 = _____

E. 180 + 27 = _____   F. 270 + 45 = _____

G. 330 − 40 = _____   H. 280 − 41 = _____

I. 445 − 50 = _____

Part 5. Multiplying by 10

1. For the following problems, make a prediction of what you think the answer will be. Then, do the problem on your calculator to check.

A. $6 \times 70 =$ _____

B. $8 \times 400 =$ _____

C. $800 \times 6 =$ _____

D. $7000 \times 4 =$ _____

E. $800 \times 8 =$ _____

F. $60 \times 4 =$ _____

2. Predict what n must be to make each number sentence true. Check your prediction on a calculator.

A. $60 \times n = 360$

B. $n \times 5 = 350$

C. $n \times 900 = 5400$

Part 6. More Multiplication

1. Solve the following problems using paper and pencil.

A.   14
       $\times 7$

B.   700
       $\times 40$

C.   42
       $\times 3$

D.   35
       $\times 8$

E.   48
       $\times 6$

F.   600
       $\times 300$

2. Use convenient numbers to estimate the following products.

A. $50 \times 61 =$

B. $89 \times 40 =$

C. $397 \times 30 =$

Name _____   Date _____

Part 7. Groups and More Groups

1. Solve the following problems using paper and pencil.

   A. 53
       ×4

   B. 38
       ×2

   C. 77
       ×3

   D. 65
       ×5

2. Bessie Coleman School is holding a fund-raiser. The money earned will go towards buying new books for the school library. Jacob is in charge of pouring lemonade at the fund-raiser. He has 19 packages of paper cups. Each package has 20 cups. About how many cups does he have?

3. Lee Yah is in charge of selling hot dogs. She has 36 packages of hot dog buns at the start of the day. Each package has 8 buns. How many hot dogs can she sell?

4. Jacob is selling raffle tickets. One raffle ticket sells for $4.

   A. So far he has collected $160. How many raffle tickets did he sell?

   B. His goal is to collect $400. How many more raffle tickets must he sell to reach his goal?

5. Ten people can sit at one table for Bingo. There are 12 tables for Bingo. How many people can play Bingo at one time?

6. At the fund-raiser, a "meal deal" that includes a hot dog, drink, and chips costs $3. There are 96 students in the eighth grade at Bessie Coleman School. If each eighth grader buys one meal deal, how much money will the eighth grade class pay in all for their food?

Triangle Flash Cards: The Last Six Facts

- Work with a partner. Each partner cuts out the 6 flash cards and lightly colors the highest number on each card.
- Your partner chooses one card at a time and covers the corner containing the highest number.
- Multiply the two uncovered numbers.
- Divide the used cards into three piles: those that you know and can answer quickly, those that you can figure out, and those that you need to learn.
- Practice the last two piles again. Then, make a list of the facts you need to practice at home.
- Repeat the directions for your partner.

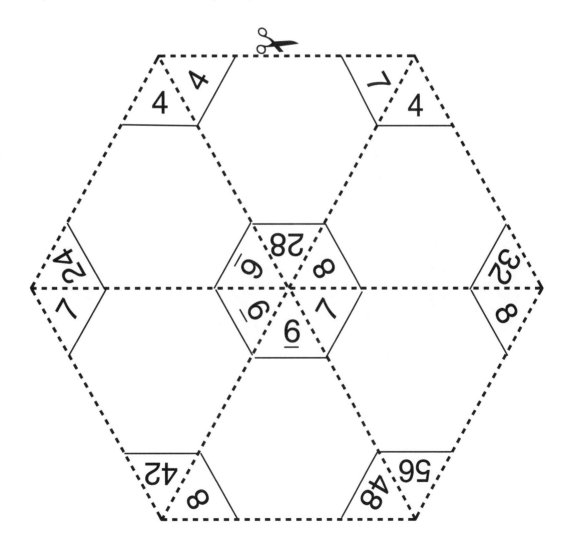

UNIT 8

Measuring Up:
An Assessment Unit

Unit 8: Home Practice

Part 1. *Triangle Flash Cards:* Reviewing All the Facts

Study for the test on all the multiplication facts. Take home your *Triangle Flash Cards* for the 5s and 10s, 2s and 3s, square numbers, 9s, and the last six facts. Study the facts in small groups, about 8–10 facts each night.

Here's how to use the flash cards. Ask a family member to choose one flash card at a time. Your partner should cover the corner containing the highest number. This number will be the answer to a multiplication fact. Multiply the two uncovered numbers.

Separate the used cards into three piles: those facts you know and can answer quickly, those that you can figure out with a strategy, and those that you need to learn. Practice the last two piles again. Remember to concentrate on one small group of facts each night—about 8 to 10 facts. Also, remember to study only those facts you cannot answer correctly and quickly.

If you do not have your flash cards, create new ones for those facts that are not yet circled on your *Multiplication Facts I Know* chart. To create your flash cards, use the *Triangle Flash Cards Master*s that follow the Home Practice in the *Discovery Assignment Book.*

Your teacher will tell you when the test on all the multiplication facts will be.

Part 2. Number Relationships

1. **A.** Is 51 prime? Tell how you know.

   B. Is 53 prime? Tell how you know.

   C. Is 55 prime? Tell how you know.

2. **A.** Is 6 a factor of 96? How can you tell?

   B. Is 6 a factor of 116? How can you tell?

3. Make a factor tree to find the prime factors of 54.

Part 3. Performing Operations

1. Solve the following problems using paper and pencil or mental math.

   A. 459 + 769 =   **B.** 1078 + 5498 =   **C.** 7089 – 2793 =

   D. 38 × 5 =   **E.** 700 × 90 =   **F.** 44 × 6 =

2. Shannon's dad is taking two night courses at a community college. Each course costs $86. Each of his two textbooks costs $45. How much does Shannon's dad have to pay to go to school?

3. In the 1994–1995 basketball season, Shaquille O'Neal of the Orlando Magic averaged 29 points a game. If Shaquille played in 79 games, about how many points did he score?

4. In the 1992 presidential election, Bill Clinton received 44,908,254 votes. George Bush received 39,102,343 votes. About how many more votes did President Clinton get than Bush?

5. It takes the Earth about 365 days to revolve around the sun. Mercury's revolution around the sun takes 277 fewer days than the Earth's. How many days does it take Mercury to revolve around the sun?

Part 4. Arithmetic Review

1. Use paper and pencil to solve the following problems. Be sure to use estimation to make sure your answers make sense.

    A. 6035 − 854 =   **B.** 27,894 + 34,676 =   **C.** 47 × 8 =

    D. 437 + 579 + 902 =   **E.** 3649 − 2089 =   **F.** 82 × 5 =

2. Use estimation to answer Questions 2A and 2B. Record number sentences to show what convenient numbers you chose. Then, solve the problem in Question 2C.

    A. Maya has $10.00 to buy the following items: cereal for $3.28, milk for $2.89, and bread for $1.56. Estimate to see if she has enough money. If there is no sales tax, about how much change will Maya receive from her $10 bill or about how much more money will she need?

    B. Shannon has been given $15.00 to buy school supplies. She wants to buy 5 books that cost $1.25 each, 10 pencils that cost 19 cents each, and 4 folders that cost 28 cents each. Estimate to see if she has enough money. About how much money will she have left? Or, how much extra will she need based on your estimation?

    C. Jerome's aunt gave him $20.00 for his birthday present. He purchased a shirt for $7.99, a cap for $2.60, and socks. After shopping, he had $5.67 left over. If no tax was paid, what was the exact cost of the socks? Look back at your answer. Is it reasonable?

Part 5. Base-Ten Shorthand

1. The bit is one whole. Label each of the following with its correct number. Then, put the numbers in order from least to greatest.

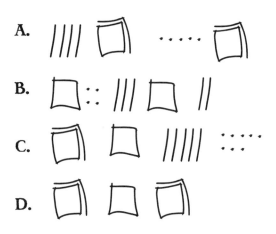

A.

B.

C.

D.

2. The following information, taken from the *1996 World Almanac,* shows the number of people employed in various occupations in the United States.

Crafts people and those who repair things: 10,435,000 people
Farming, forestry, and fishing: 1,430,000 people
Machine operators and truck drivers: 14,001,000 people
Professionals (accountants, doctors, lawyers,
    teachers, nurses):   23,247,000 people
Sales people, technicians, administrative workers: 25,928,000 people
Service jobs such as waiters, guards, and janitors: 9,104,000 people

A. List the number of people employed in the different kinds of jobs in order from least to greatest.

B. About 3,383,000 of the machine operators and truck drivers are women. About how many are men?

C. About 12,082,000 of the professionals are men. About how many are women?

Triangle Flash Cards Master

- Make a flash card for each fact that is not circled on your *Multiplication Facts I Know* chart.
- Work with a partner. Each partner cuts out the flash cards and lightly colors the highest number on each card.
- Your partner chooses one card at a time and covers the corner containing the highest number.
- Multiply the two uncovered numbers.
- Repeat the directions for your partner.
- Study the facts you need to learn in small groups, about 8–10 facts a night.

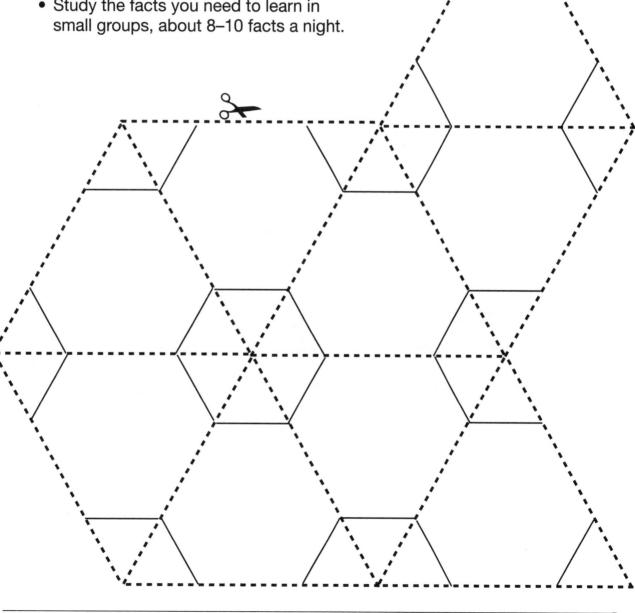

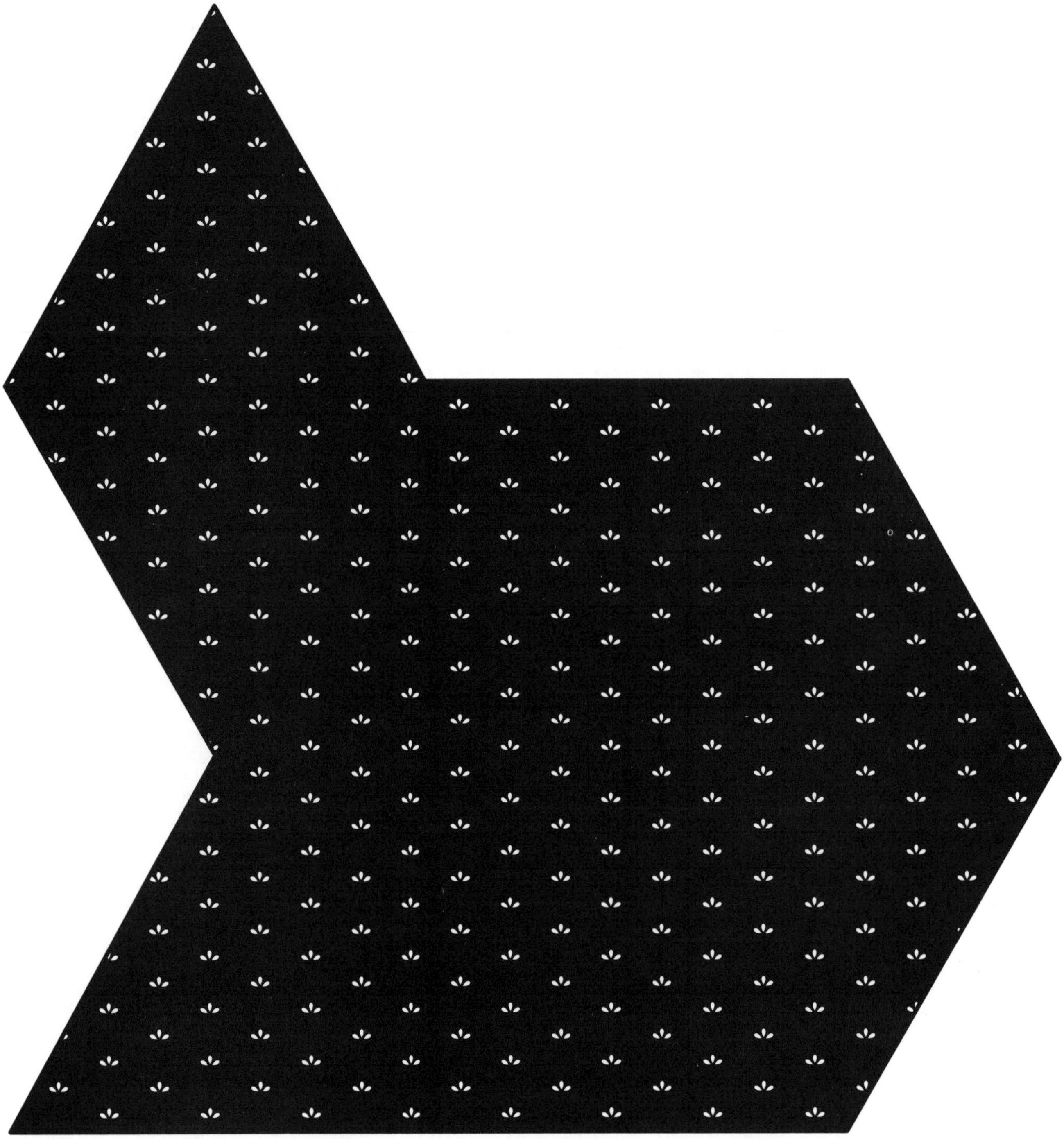

Triangle Flash Cards Master

- Make a flash card for each fact that is not circled on your *Multiplication Facts I Know* chart.
- Work with a partner. Each partner cuts out the flash cards and lightly colors the highest number on each card.
- Your partner chooses one card at a time and covers the corner containing the highest number.
- Multiply the two uncovered numbers.
- Repeat the directions for your partner.
- Study the facts you need to learn in small groups, about 8–10 facts a night.

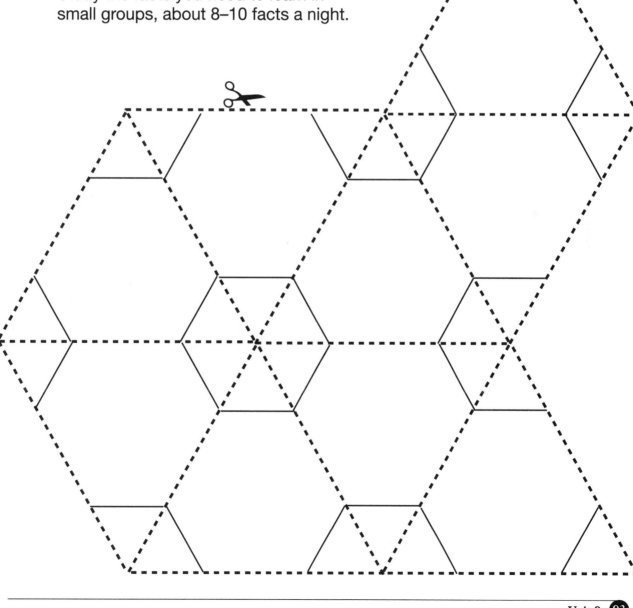

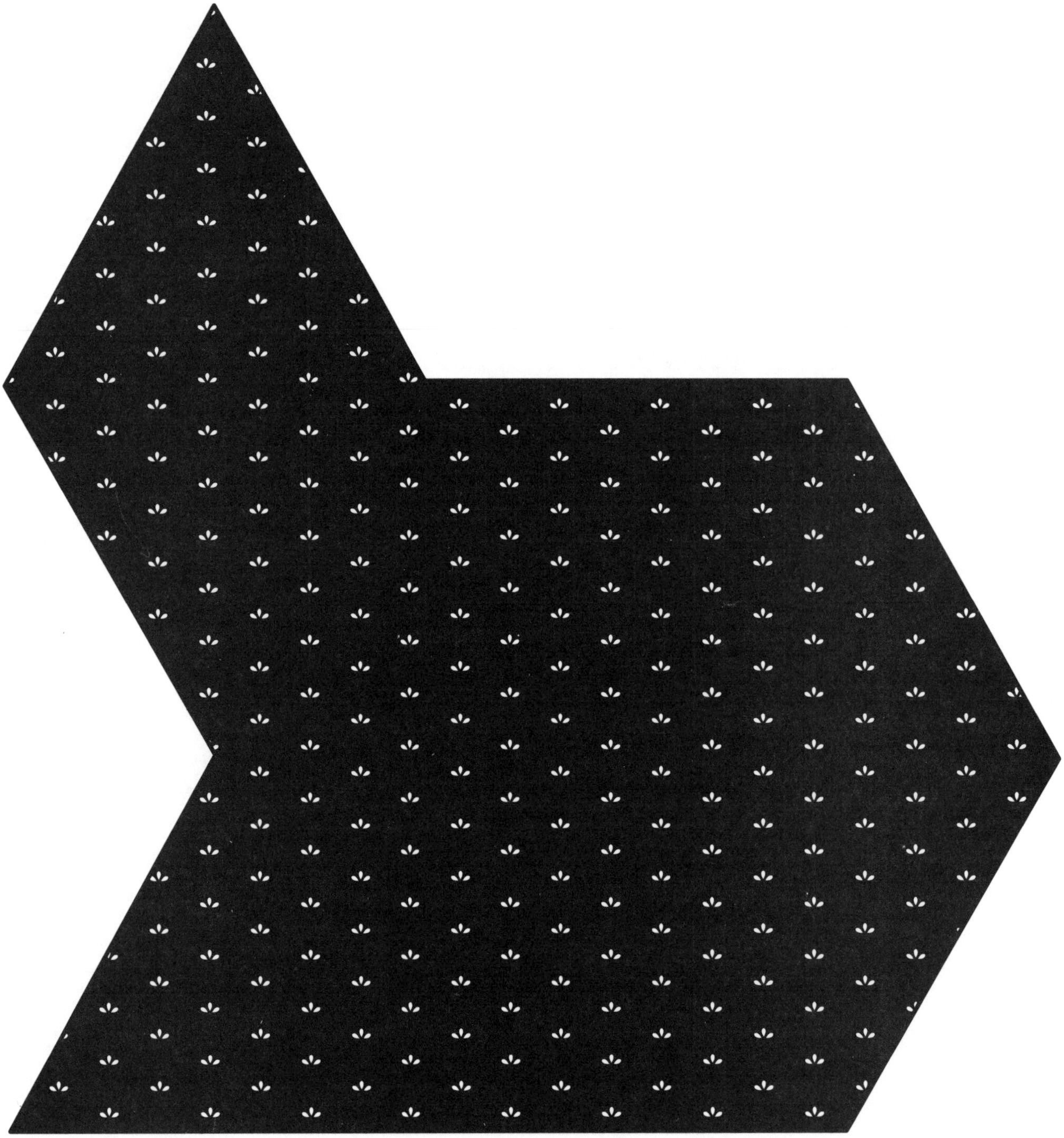

Name _____ Date _____

Estimating and Measuring Volume

1. Use 8 centimeter connecting cubes
   to make an object that will fit into
   a 250-cc graduated cylinder.
   What is the volume of your object?

   A. Fill a 250-cc graduated cylinder
      with a convenient amount of
      water. Good choices are
      160 or 200 cc. Use an
      eyedropper to carefully
      add the last few drops.

   B. To read the water level correctly, put your eyes at the level of the water.
      When water creeps up the sides of a cylinder, it forms a **meniscus** which
      makes it look as though there are two lines. Read the lower line.

   C. Place your object made from connecting cubes into the cylinder. Slide it
      in gently so that no water will splash. Read the water level now.

   D. What is the difference in the level of the water before you added the
      object and after you added it? Explain the change in water level.

One way to find the volume of an object is to put the object underwater in a
graduated cylinder and see how much the water rises. This is called finding **volume
by displacement**. You are going to estimate the volume of objects using centimeter
connecting cubes and find the volume of these objects by displacement.

2. Choose objects that
   will fit into a graduated
   cylinder. (Your teacher
   will help you.) Make a
   model of your objects
   using centimeter
   connecting cubes.
   Estimate the volume
   of the objects by
   counting the
   number of cubes
   in your models.

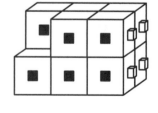

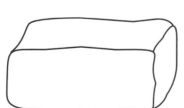

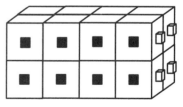

3. Find the volume of your objects by displacement.

4. Record your results in a table like the one below. Follow the examples.

Volume Data Table

Object	Estimated Volume from Cube Model	Volume by Displacement
Rock	11 cc	12 cc
Marker	14 cc	11 cc
Clay	16 cc	15 cc

5. Frank made a model of a marker using centimeter connecting cubes. By counting the cubes, he estimated that the marker has a volume of 14 cc. When he measured the volume using a graduated cylinder, he found the volume to be 11 cc. Why do you think there is a 3 cc difference?

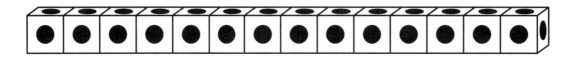

6. Were your estimates close to your measured volumes? Why or why not?

Homework

1. Look at the two scales on this page. How are they alike? How are they different?

2. How many cubic centimeters (cc) does each space represent on the scale for a 100-cc graduated cylinder?

3. How many cubic centimeters does each space represent on the scale for the 250-cc graduated cylinder?

4. Write the scale reading for each letter.

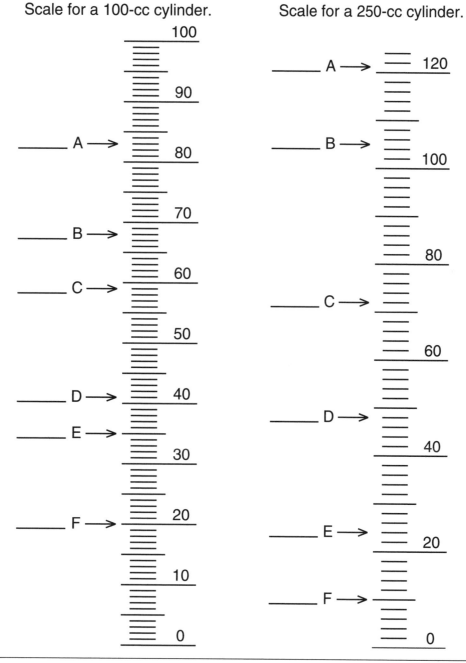

Scale for a 100-cc cylinder.

Scale for a 250-cc cylinder.

5. Shannon put a piece of clay under water in a graduated cylinder. What was the volume of the clay?

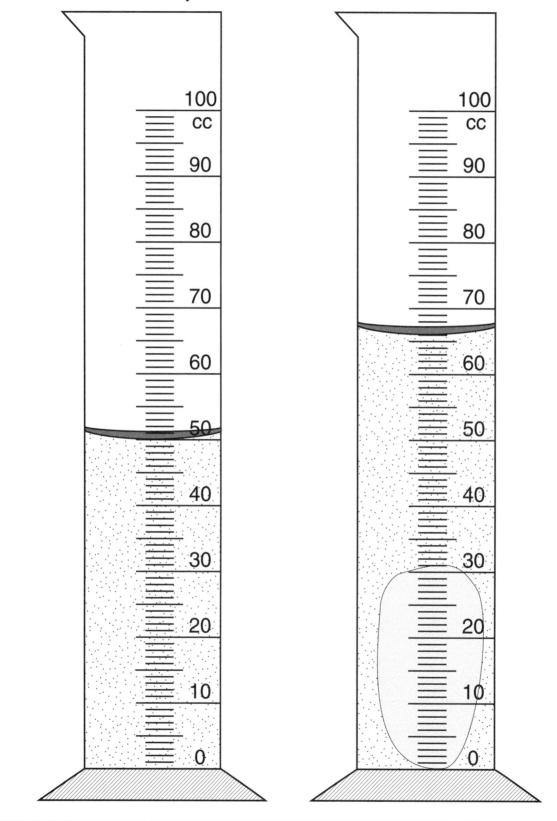

Mixed-Up Multiplication Tables

Homework

Fill in these multiplication tables:

1.

×	2	4	8
2			
4		16	
8			

2.

×	1	3	9
1			
3			
9			

3. What patterns do you see in the table in Question 1?

4. What patterns do you see in the table in Question 2?

5.

×	10	5	0
10			
5			
0			

6.

×	6	5	7
6			
5			
7			

7. What patterns do you see in the table in Question 5?

8. What patterns do you see in the table in Question 6?

9.

×	8	6	4
8			
6			
4			

10.

×	8	6	3
8			
6			
3			

11.

×	2	5	8
9			
4			
7			

12.

×	7	6	4
3			
9			
10			

Spinner 1–4

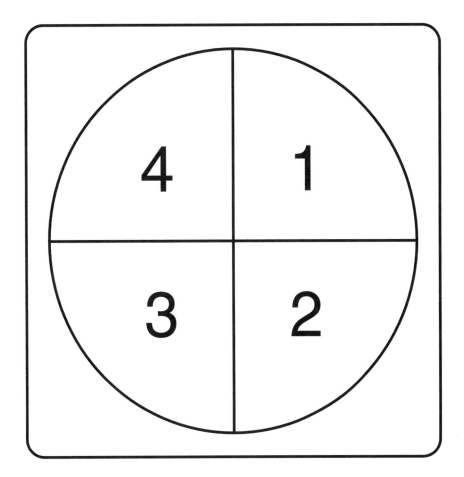

Name _____ Date _____

Predicting Prices

Homework

Look for patterns in each graph. If the points form a line, draw a best-fit line on the graph.

1.

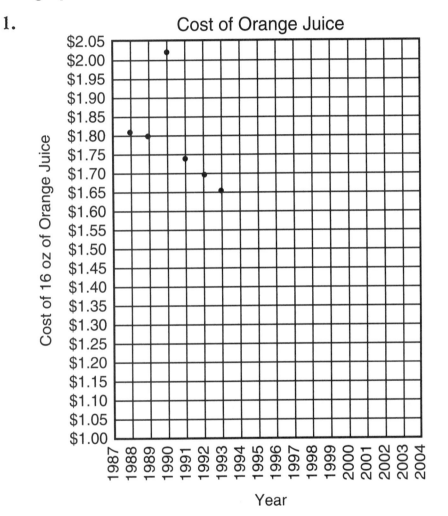

Cost of Orange Juice

A. Using this data, can you make a reasonable prediction about the cost of orange juice in 1999? Why or why not?

B. If possible, predict the cost of 16 oz of orange juice in the year 1999.

C. Describe the graph. What do you think happened to the orange crop in 1990?

2.

Cost of Lettuce

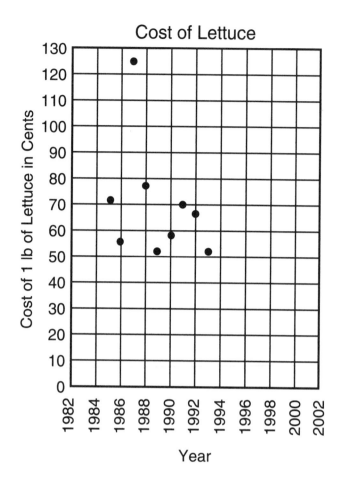

Year

A. Using this data, can you make a reasonable prediction about the cost of a pound of lettuce in the year 2000? Why or why not?

B. If possible, predict the cost of 1 pound of lettuce in the year 2000.

Volume vs. Number

3.

Balloon Rides

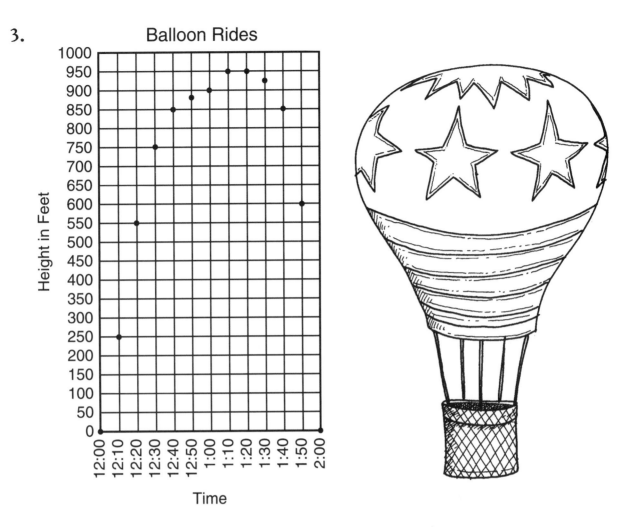

A. Using this data, can you make a reasonable prediction about the height of the balloon at 12:15? Why or why not?

B. If possible, predict the height of the balloon at 1:45.

C. Describe the graph. Why do you think the graph has the shape it does?

Experiment Review Chart

Directions:
- Write the names of the experiments completed this year in the first row of the table.
- Complete each column with information for each lab.

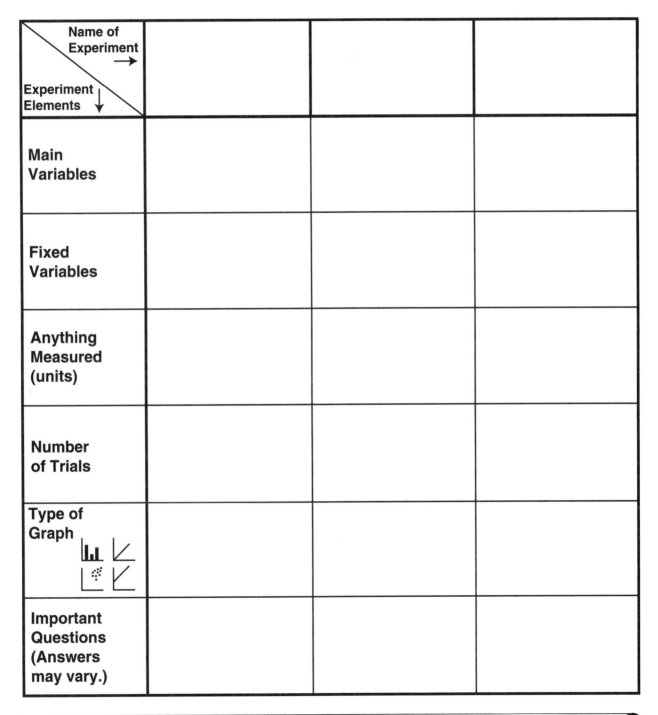

Name _____   Date _____

Directions:

- Write the names of the experiments completed this year in the first row of the table.
- Complete each column with information for each lab.

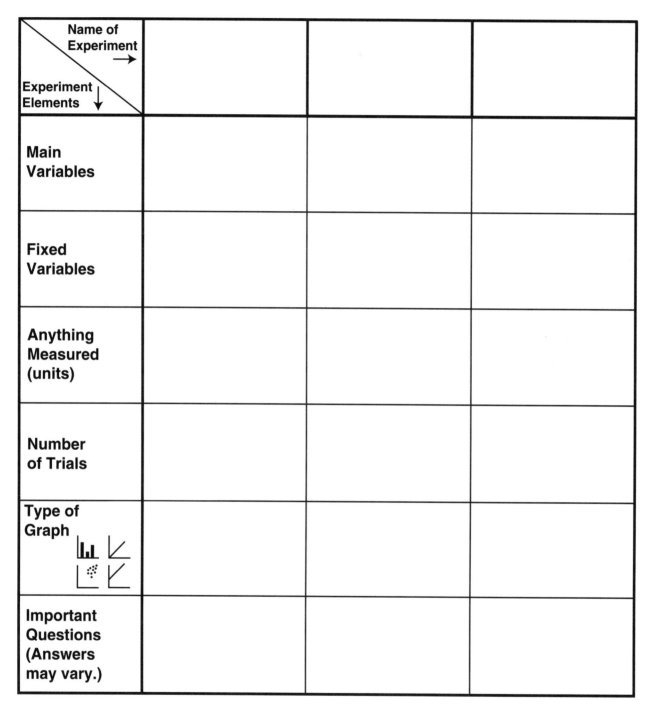

Name of Experiment → Experiment Elements ↓			
Main Variables			
Fixed Variables			
Anything Measured (units)			
Number of Trials			
Type of Graph			
Important Questions (Answers may vary.)			

Mid-Year Experiment and Portfolio Review

UNIT 9

Shapes and Solids

Unit 9: Home Practice

Part 1. *Triangle Flash Cards:* 5s and 10s and the Square Numbers

Study for the quiz on the multiplication facts for the 5s and 10s and the square numbers. Take home your *Triangle Flash Cards* and your list of facts you need to study.

Here's how to use the flash cards. Ask a family member to choose one flash card at a time. Your partner should cover the corner containing the highest number. This number will be the answer to a multiplication fact. Multiply the two uncovered numbers.

Your teacher will tell you when the quiz on the 5s and 10s and the square numbers will be. Remember to concentrate on one small group of facts each night—about 8 to 10 facts. Also, remember to study only those facts you cannot answer correctly and quickly.

Part 2. Mental Multiplication

1. Solve Questions 1A–1F in your head.

   A. $6000 \times 10 =$ _____

   B. $500 \times 70 =$ _____

   C. $60 \times 60 =$ _____

   D. $9000 \times 5 =$ _____

   E. $100 \times 800 =$ _____

   F. $500 \times 50 =$ _____

2. A. Paper towels cost 80¢. How much will 6 rolls cost?

   B. An ice cream bar costs $3.50 each. How much will 4 bars cost?

   C. Bagels cost 69¢ each. About how much will 5 bagels cost?

3. How much is:
   A. 13 nickels?

   B. 11 nickels and 5 dimes?

   C. 5 quarters and 17 nickels?

Part 3. Addition and Subtraction

1. Find the missing numbers needed to make these addition and subtraction problems correct. Use pencil and paper only.

A.   189
     + ___
      612

B.   322
     − ___
      284

C.   5078
     + ___
      8079

D.   7339
     − ___
      6079

E.   5405
     + _____
      13,053

F.   3000
     − ___
      1456

2. Estimate the following products using convenient numbers. Write a number sentence to show your thinking.

A. $290 \times 18 =$

B. $505 \times 59 =$

C. $9956 \times 9 =$

3. Find the products using paper and pencil. Be sure to estimate to make sure your answers are reasonable.

A. $63 \times 4 =$

B. $37 \times 8 =$

C. $28 \times 9 =$

D. $84 \times 4 =$

E. $66 \times 3 =$

F. $72 \times 6 =$

Part 4. Time

1. What time is shown on each clock below?

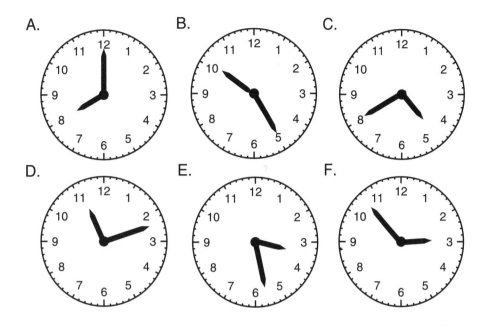

A.   B.   C.

D.   E.   F.

2. **A.** Jackie began cleaning her room at 6:15 P.M. If it took Jackie two hours to clean her room, at what time did she finish?

B. John started watching cartoons when he woke up 1 hour and 15 minutes ago. It is now 10:00 A.M. What time did John wake up?

C. Jacob's mother has to pick him up from his aunt's house at 9:30 P.M. It takes 50 minutes to get there. What time should she leave her home?

D. Irma and her sister are making dinner for the family. They plan to eat at 6:30 P.M. The dinner takes 2 hours and 35 minutes to prepare. What time should they begin cooking?

E. How many minutes will it take Irma and her sister to prepare dinner?

F. How many hours is 600 minutes?

Part 5. Geometry

You will need a protractor, a ruler, and two pieces of *Centimeter Grid Paper* to complete this part of the Home Practice.

1. Nicholas built the prism at the right from centimeter connecting cubes.

   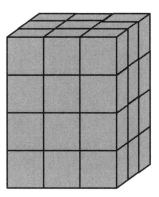

   A. How many centimeters tall is the prism?

   B. How many square centimeters cover the base of the prism?

   C. What is the volume of the prism?

2. What is the volume of each of the prisms below? Show how you found your answer.

   A.

   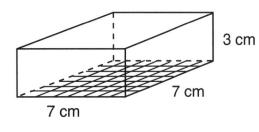

   3 cm

   7 cm

   7 cm

   B. The area of the base is 56 sq cm.

   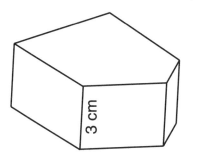

   3 cm

3. Go to a cupboard (ask an adult first). Select three objects that are prisms (boxes) such as a gelatin box, a small cereal box, and a toothpaste box.

   A. Predict which box has the greatest volume and which has the least volume.

   B. Trace the base of each on *Centimeter Grid Paper*. How many sq cm cover the base of each box?

   C. Measure the height of each box in cm.

   D. Use the information you gathered in 3B and 3C to find the volume of each container. Were your predictions close to your estimates?

4. Draw each of the angles described below with a ruler and protractor on a separate sheet of paper.

   A. Draw a 35° angle. Name the angle ∠QRT.

   B. Draw a 142° angle. Name the angle ∠XYZ.

Measuring Angles I

Measure each of the angles in Questions 1–6 with your protractor. To measure, you may need to extend the sides of some of the angles with the edge of your protractor. Label each angle as acute, obtuse, or right.

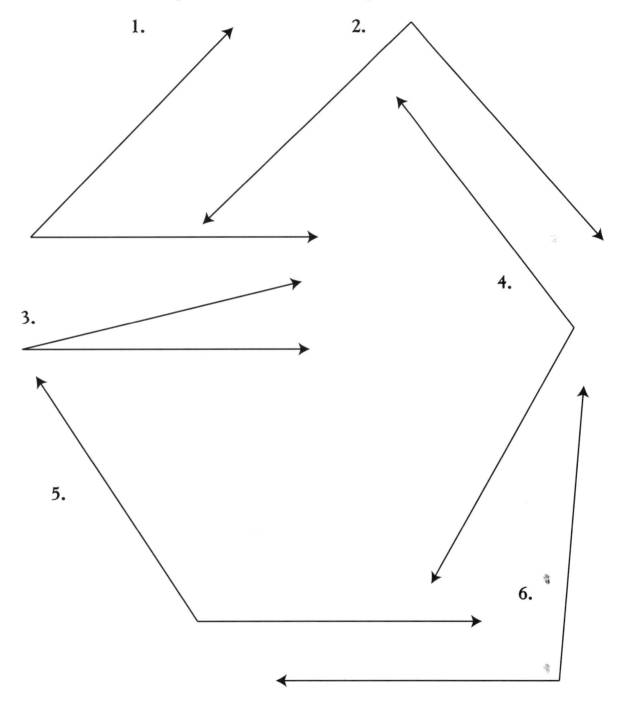

Measuring Angles II

Measure each of the angles in Questions 1–6 with your protractor. You may need to extend the sides of some of the angles in order to measure. Label each angle as acute, obtuse, or right.

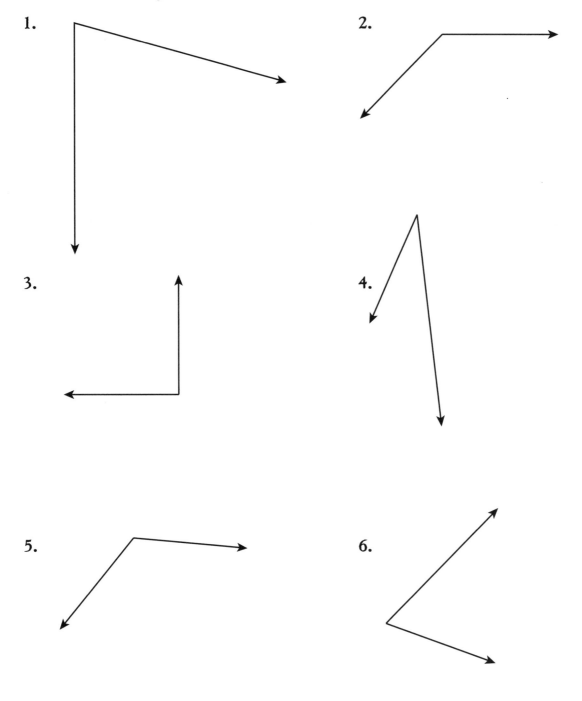

1.

2.

3.

4.

5.

6.

Pattern Block Shapes

Cut out the shapes below. Answer the following questions for each. Does the shape have line symmetry? If the shape has line symmetry, how many lines of symmetry does it have? Draw in the lines of symmetry.

Turn Symmetry Examples

The pictures here all have turn symmetry. The center of turning is marked on each.

1. Cut out each of the shapes on the bottom half of this page.

2. Place each cutout on top of its template below. Place your pencil point directly on the center of turning. As you turn the cutout 360° (a full turn), how many times do the two pictures match after turning?

3. Find the type of turn symmetry for each picture. Label each one.

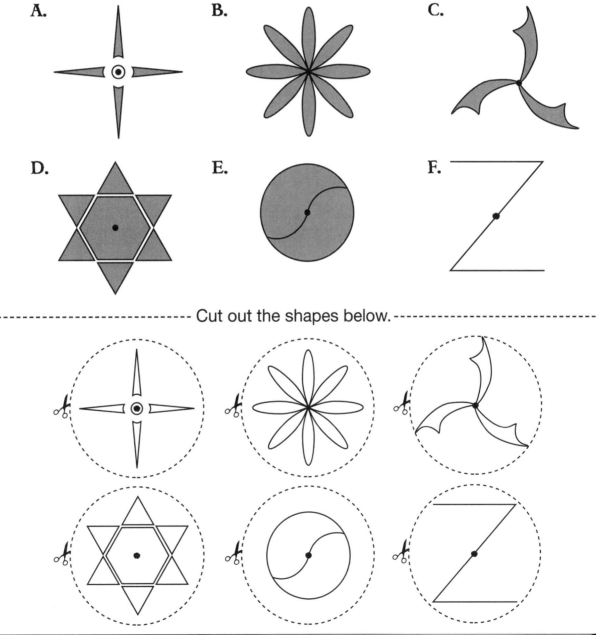

A.   B.   C.

D.   E.   F.

---------------------------- Cut out the shapes below. ----------------------------

Line Symmetry Examples

Each of the pictures here has line symmetry.

1. Cut out each shape.

2. Fold the shapes to find the line(s) of symmetry. Remember, if you fold the shape along a line of symmetry, the two pieces will fit exactly on top of one another.

3. Draw a line(s) of symmetry on each shape.

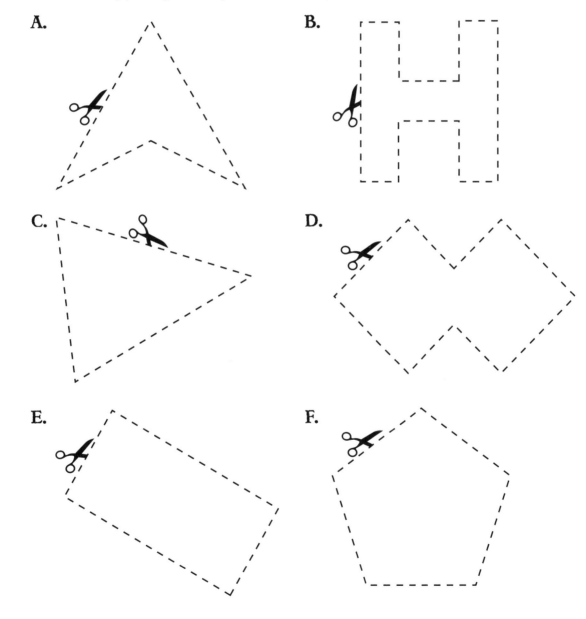

A.
B.
C.
D.
E.
F.

Name _____   Date _____

More Symmetry
Homework

1. Parts of the pictures below are missing. The lines are lines of symmetry. Draw in the missing parts.

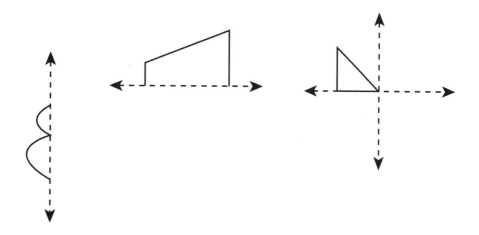

2. Which of the pictures below have turn symmetry? What kind of turn symmetry? (*Hint:* Trace the shapes and cut out the tracings. Then, place the tracing on top of the picture and turn it.)

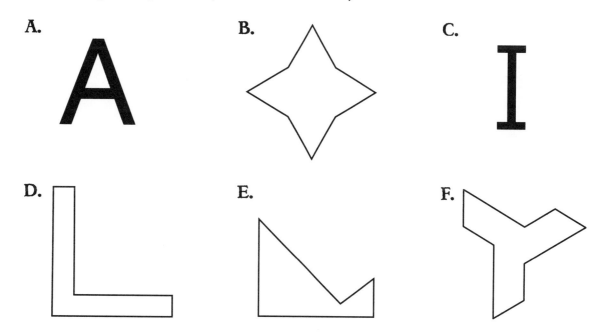

A.   B.   C.

D.   E.   F.

3. Which of the pictures in Question 2 have line symmetry? Draw in the line(s) of symmetry. (*Hint:* You may cut out the shapes and fold them.)

Blank Spinners

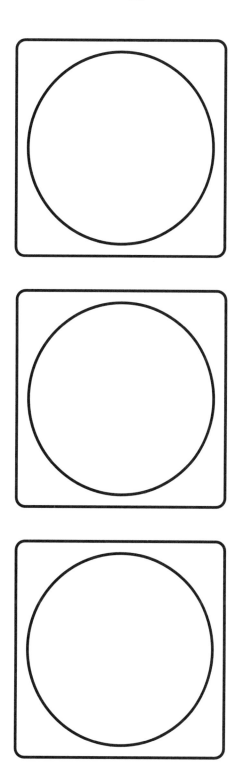

Nets

Here is a net of a cube.

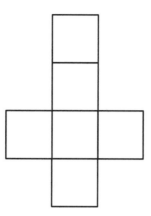

Which of the following figures is a net of a cube? You may wish to cut out the figures to help you decide. Cut on the dotted lines and fold on the solid lines.

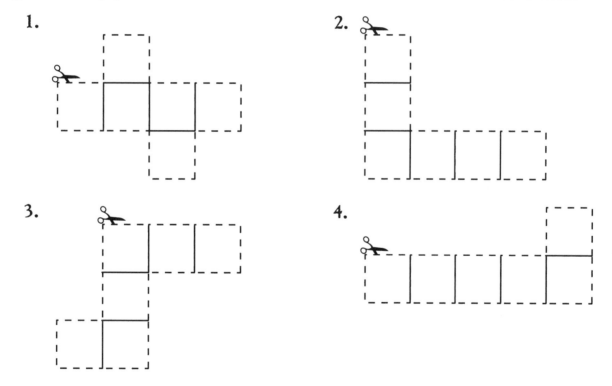

1.

2.

3.

4.

5.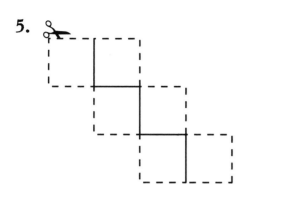

6.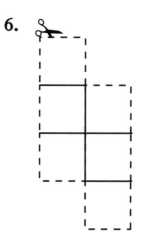

A Hexagonal Prism

You will need a protractor, ruler, scissors, and tape to complete the activity. Answer Questions 1–6 before cutting out the net of the hexagonal prism.

Look at the net on the following page. It contains two hexagons. Hexagons are polygons with six sides.

1. What is the measure of the angles of the hexagons?

2. What is the length of the sides of the hexagons in centimeters?

3. What are the lengths of the sides of the rectangles in centimeters?

4. What point on the rectangle will point E touch when the net is cut and folded?

5. What point on the rectangle will point D touch when the net is cut and folded?

6. What point on the rectangle will point B touch when the net is cut and folded?

Carefully cut the net. If you are using heavier paper, score the inside lines. To score means to use a ruler and sharp pencil to trace the lines you are going to fold. It will make the folds much neater. Then, fold and tape.

You may wish to decorate the net before cutting and folding. Think about which side will be on the inside and which side will show on the outside.

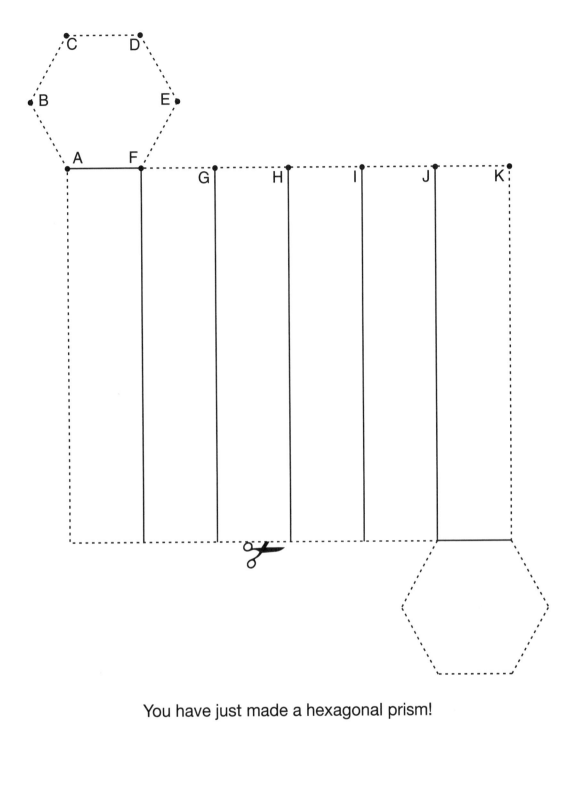

You have just made a hexagonal prism!

Right Triangular Prism

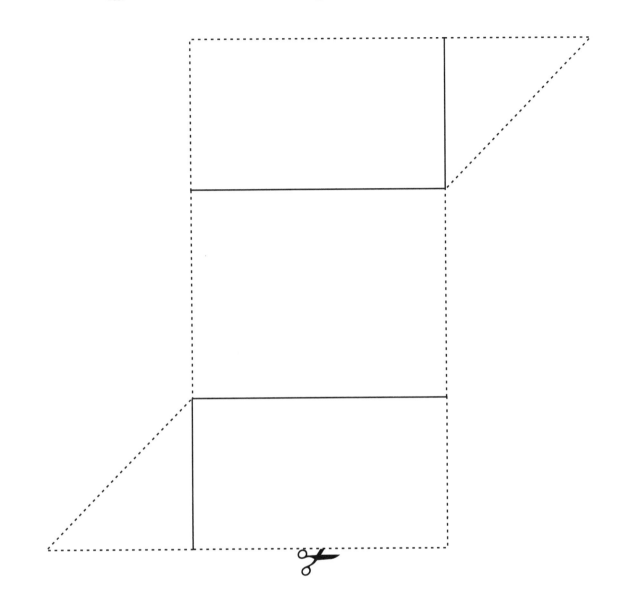

Net of an Octahedron

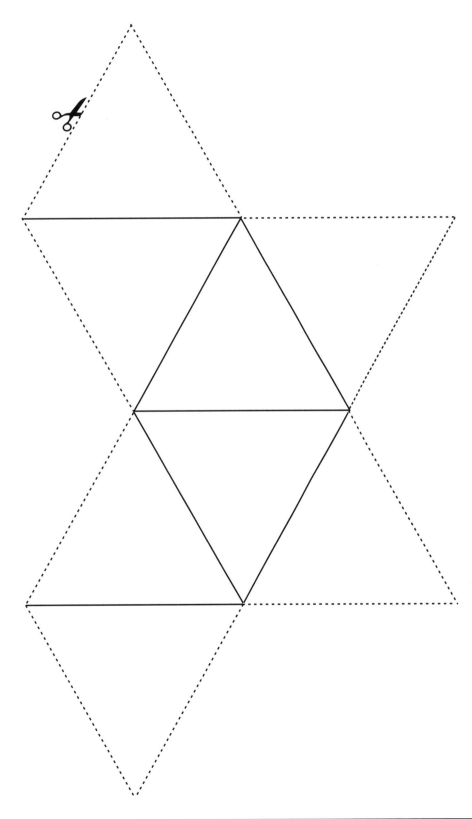

UNIT 10

Using Decimals

Unit 10: Home Practice

Part 1. *Triangle Flash Cards:* 2s, 3s, and the 9s

Study for the quiz on the multiplication facts for the 2s, 3s, and the 9s. Take home your *Triangle Flash Cards* and your list of facts you need to study.

Here's how to use the flash cards. Ask a family member to choose one flash card at a time. Your partner should cover the corner containing the highest number. This number will be the answer to a multiplication fact. Multiply the two uncovered numbers.

Your teacher will tell you when the quiz on the 2s, 3s, and the 9s will be. Remember to concentrate on one small group of facts each night—about 8 to 10 facts. Also, remember to study only those facts you cannot answer correctly and quickly.

Part 2. Missing Numbers and Big Numbers

1. What number must *n* be to make each number sentence true. After you have decided on a number for *n,* check your work by multiplying.

   A. $n \times 20 = 80$ **B.** $300 \times n = 1800$ **C.** $90 \times n = 5400$

   D. $50 \times n = 10,000$ **E.** $n \times 90 = 8100$ **F.** $70 \times n = 210$

2. **A.** Write the following numbers in order from smallest to largest.

   45,676   54,673   45,788   48,654   47,998   45,089

   B. Round 48,654 to the nearest thousand.

   C. Round 45,089 to the nearest hundred.

3. Use convenient numbers to estimate the answers to the following problems. Record number sentences to show your thinking.

   A. $608,965 + 28,696$ **B.** $2,657,223 + 3,908,700$ **C.** $378,904 - 99,645$

Part 3. Decimals

1. Numbers are represented in base-ten shorthand below. The flat is one whole. Label each of the following with its correct number. Then, put the numbers in order from least to greatest.

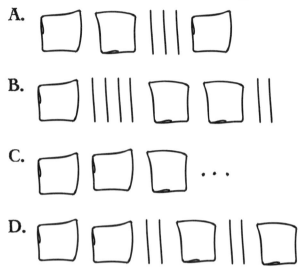

A.

B.

C.

D.

2. Write a decimal for each of the following. Then, write your decimal using base-ten shorthand. The flat is one whole. Find a number that is:

   A. Between 8 and 9   B. Between 4 and 4.5

   C. Just a little bigger than 8   D. Between $\frac{1}{2}$ and 2

3. For each of the two problems below, put the measurements in order from shortest to longest.

   A. 0.6 m   23 cm   1 dm   0.45 m   55 dm

   B. 1.5 m   1 m and 8 dm   1.03 meter   1.24 meter

Part 4. School Supplies

Linda and her brother are buying school supplies. Notebooks are on sale for 39¢ each. Pencils are 4 for $1.00. A set of markers costs $2.98. Folders are 10 for $1.00.

1. Linda needs 3 notebooks, 1 set of markers, 1 folder, and 8 pencils. Estimate the cost of Linda's school supplies. Use number sentences to show your thinking.

2. Linda's brother needs 5 notebooks, 1 set of markers, 3 folders, and 4 pencils. Estimate the cost of his school supplies. Use number sentences to show your thinking.

3. What is the exact cost of each child's supplies? (There is no tax.)

4. What is the difference in price between the two children's supplies? Use a number sentence to show how you solved the problem.

Part 5. Addition, Subtraction, and Multiplication

Solve the following problems using paper and pencil. Estimate to make sure your answers are reasonable.

A. $68 - 49 =$ **B.** $167 + 67 =$ **C.** $284 + 238 =$ **D.** $432 - 197 =$

E. $47 \times 9 =$ **F.** $26 \times 7 =$ **G.** $34 \times 9 =$ **H.** $23 \times 8 =$

Part 6. Playing at the Park

1. **A.** When Shannon and her family arrived at the park on Saturday, Shannon counted 3 children on *each* of the following: the slide, the swings, the monkey bars, and the merry-go-round. How many children were at the park when Shannon arrived?

   B. If there were 8 more children than adults at the park, how many adults were at the park?

2. A used car dealer is across the street from the park. Shannon's dad looked at some cars while Shannon and her sister played at the park. He liked two different cars. One car costs $4550 and the other costs $3775. What is the difference in price of the two cars?

3. Shannon treated her little sister and her mother to a treat. At a nearby stand she bought two cans of juice at 65¢ each and three popsicles at 85¢ each. She gave the vender $5.00. How much change will Shannon receive?

4. While playing in the park, Shannon's family saw a 5 kilometer race. 235 people were signed up to participate, but only 178 arrived the day of the race. How many people did not show up for the race?

5. **A.** Last summer, the park district raised money for new playground equipment. In June, $565 was raised. In July, $438 was raised. In August, $395 was raised. How much money was raised for new playground equipment?

   B. How much money do they need to raise in September to reach their goal of $1500?

Measure Hunt

Find four lengths in the classroom using the rules in the table below. Use the fewest pieces possible when you measure. Use metersticks to measure meters. Use skinnies to measure decimeters. Use bits to measure centimeters. Then, complete the table. An example is done for you.

Table 1

Rule	Object	Number of			Length (nearest 0.01 m)
		m	**dm**	**cm**	
Between 1 and 2 m	Height of cabinet	1	2	0	1.20
Between 1 and 1.5 m					
Between 0.5 and 1 m					
Between 0.5 and 0.8 m					
Between 0.55 and 0.75 m					

Find five lengths in the classroom using the rules in the table below. Use metersticks to measure the lengths. Then, complete the table.

Table 2

Rule	Object	Number of			Length (nearest 0.01 m)
		m	**dm**	**cm**	
Between 1 and 2 m					
Between 1 and 1.5 m					
Between 0.5 and 1 m					
Between 0.6 and 0.9 m					
Between 0.40 and 0.65 m					

m, dm, cm, mm

Tenths Helper

Place whole skinnies on this chart to find how many tenths are in one whole and two wholes. Skip count by tenths.

Exploring Tenths

1. Michael placed 14 skinnies on his *Tenths Helper* chart. Write the number this represents in more than one way.

2. **A.** Grace wanted to build the number 1.9 on her *Tenths Helper* chart using skinnies. How many skinnies will she need to build this number?

   B. Grace decided to use both flats and skinnies to build this number. How many flats will she need? How many skinnies?

3. Nila built the following number on her *Tenths Helper* chart:

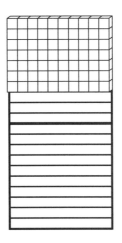

   Write this number in more than one way.

4. **A.** Jackie placed 4 flats on her desk. How many whole units does this represent?

   B. She added 6 skinnies to the 4 flats. What number does 4 flats and 6 skinnies represent?

5. **A.** Jacob placed 6 skinnies along the edge of a meterstick. The length of 6 skinnies is what fraction of a meter?

   B. Write this fraction as both a decimal fraction and a common fraction.

6. **A.** Jessie measured the length of a table in her classroom using skinnies. She found the length of the table was 18 skinnies. How many whole meters and how many more tenths does this represent?

   B. Write the length of the table in more than one way.

7. Use what you know about tenths to complete the table by filling in the missing information. (Remember: A flat is one whole.)

Base-Ten Shorthand	Common Fraction	Decimal Fraction
||||| ||||	$\frac{9}{10}$	
		6.7
▢▢▢▢▢▢||||		
	$23\frac{6}{10}$	
		52.3
▢▢▢▢||||| ||		
▢▢▢▢ ||		
	$34\frac{5}{10}$	
▢▢▢▢▢|||		

Grace's Base-Ten Pieces

You need base-ten pieces (packs, flats, and skinnies only). A flat is one whole.

1. If a flat is 1, then what number is a pack?

2. If a flat is 1, then what number is a skinny?

Grace has two base-ten pieces. She might have skinnies, flats, or packs. For example, she might have a skinny and a flat. She might have something else.

3. Find all possible sets of pieces that Grace might have. Use base-ten shorthand to show each set she might have. Write a number for each set.

4. What is the largest number that Grace could possibly have?

5. What is the smallest number that Grace could possibly have?

6. Put the numbers that Grace could have in order from smallest to largest.

More Hundredths

Making a Hundredths Chart

Professor Peabody made a hundredths chart. He forgot to fill in some of the chart. Help Professor Peabody by filling in the missing values.

0.01	0.02				0.06				0.1
	0.12			0.15			0.18		
0.21			0.24			0.27			
0.31		0.33			0.36				0.4
				0.45			0.48		
	0.52								
0.61				0.65				0.69	0.7
		0.73			0.76			0.79	
0.81							0.88		0.9
0.91		0.93				0.97			1

Use your completed chart to answer the following questions.

1. **A.** What number comes after 0.09?

   B. Why is it recorded as 0.1?

2. What number comes after 0.99?

3. Describe any patterns that you see in your hundredths chart.

4. Use base-ten shorthand to make these numbers.

    A. 19.06

    B. 0.68

    C. 1.73

5. Give a decimal fraction and a common fraction for the base-ten shorthand below:

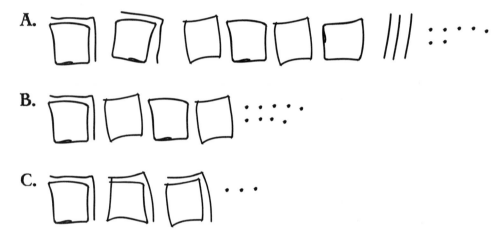

Hundredths, Hundredths, Hundredths

This is a game for two people. You need base-ten pieces (flats, skinnies, and bits). Use the *Shorthand and Fractions Table* to record your work. For this game, a flat is one whole.

Shorthand and Fractions Table

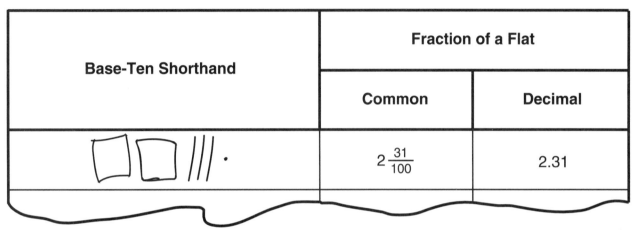

Base-Ten Shorthand	Fraction of a Flat	
	Common	**Decimal**
	$2\frac{31}{100}$	2.31

To play, the first player makes a number with base-ten pieces and shows it in the table using base-ten shorthand.

Then, the second player must write the fractions for the number and say the number.

The second player scores one point for writing the common fraction correctly, one point for writing the decimal fraction correctly, and one point for saying the number correctly.

Take turns making, writing, and saying numbers.

In this game, you are allowed to be tricky. Can you tell what this number is?

Hundredths

Shorthand and Fractions Table

You need base-ten pieces. For this page, a flat is one whole.

Base-Ten Shorthand	Fraction of a Flat	
	Common	Decimal

Decimal Hex

This is a game for two or three players.

Materials

- *Decimal Hex* Game Board on the following page
- two same color centimeter cubes or other game markers for each player
- one clear plastic spinner or pencil and paper clip

Rules

The goal of this game is to move two cubes or other markers from matching hexagons to opposite matching hexagons that all have the same number.

1. Each player places both of his or her cubes on two matching hexagons with the same number. The target hexagons are the matching ones with the same number on the other side of the game board.

2. The first player spins the spinner.

3. If "Greater Than or Equal To" shows, the player can move one cube to a neighboring hexagon with a number that is greater than or equal to the number in the hexagon where the cube is now.

4. If "Less Than or Equal To" shows, the player can move one cube to a neighboring hexagon with a number that is less than or equal to the number in the hexagon where the cube is now.

5. The player does not have to move a cube during his or her turn.

6. More than one cube can be on the same hexagon at the same time.

7. Players take turns spinning the spinner and moving cubes.

8. The first player to get both cubes to his or her target hexagons is the winner.

Name _____   Date _____

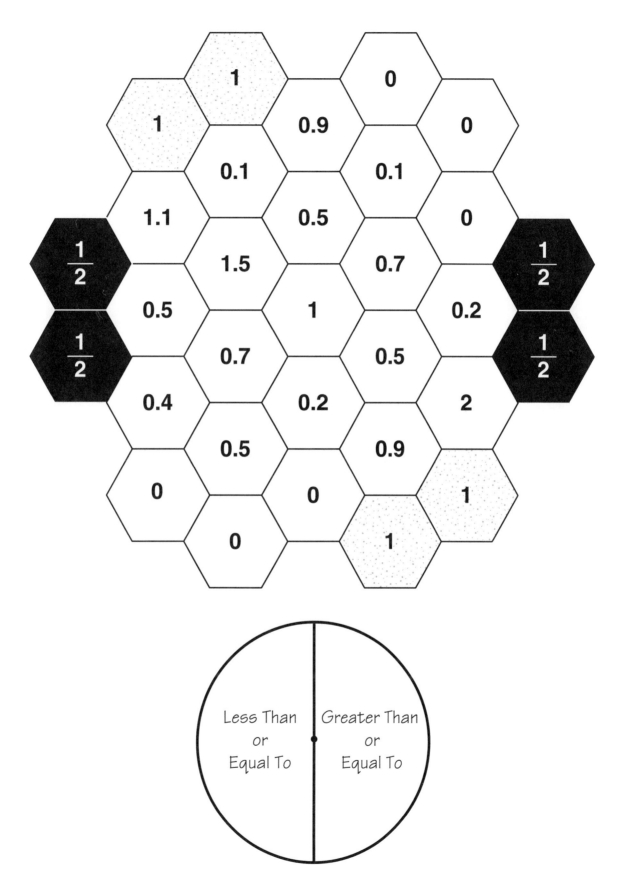

UNIT
11

Multiplication

Unit 11: Home Practice

Part 1. *Triangle Flash Cards:* The Last Six Facts

Study for the quiz on the multiplication facts for the last six facts (4×6, 4×7, 4×8, 6×7, 6×8, and 7×8). Take home your *Triangle Flash Cards* and your list of facts you need to study.

Here's how to use the flash cards. Ask a family member to choose one flash card at a time. Your partner should cover the corner containing the highest number. This number will be the answer to a multiplication fact. Multiply the two uncovered numbers.

Your teacher will tell you when the quiz on the last six facts will be. Remember to study only those facts you cannot answer correctly and quickly.

Part 2. Big Numbers

1. Write the following numbers:

   A. six hundred thirty thousand _____

   B. one million, four hundred ten thousand, nineteen _____

2. Write the following in words:

   A. 420,079 _____

   _____

   B. 6,122,038

   _____

   _____

3. Choose from the following numbers to answer Questions 3A and 3B.

   46,998   56,888   45,788   48,998   45,088

   A. If you add 2 to one of the numbers, you will get 49,000. Which number?

   _____

   B. If you add 11,100 to one of the numbers, you will get 58,098. Which number?

   _____

Part 3. Multiples of 10 and 100

Do the following problems in your head. Write only the answers.

A. $9 \times 30 =$ _____   **B.** $20 \times 30 =$ _____   **C.** $90 \times 20 =$ _____

D. $900 \times 300 =$ _____   **E.** $60 \times 90 =$ _____   **F.** $40 \times 800 =$ _____

G. $70 \times 600 =$ _____   **H.** $30 \times 800 =$ _____   **I.** $5 \times 30 =$ _____

Part 4. Writing Numbers

For each of the following numbers, write two other numbers: one that is a little smaller and one that is a little larger. You may use a number line to help you.

	A Little Smaller	A Little Larger
A. one million	_____	_____
B. one-half million	_____	_____
C. ten million	_____	_____
D. 1,300,000	_____	_____
E. 999,999	_____	_____
F. five thousand	_____	_____

Part 5. Addition and Subtraction Practice

Use paper and pencil only. Estimate to see if your answers are reasonable.

A.   92
    $+ 65$

B.   340
    $- 235$

C.   58
    $- 49$

D.   5001
    $- \ 999$

E.   1490
    $+ \ 453$

F.   289
    $+ \ 39$

Part 6. More Multiples of 10 and 100

Use paper and pencil only to solve the following problems. Show your work on a separate sheet of paper. Estimate to see if your answers are reasonable. Be prepared to explain your estimates.

A. $25 \times 20 =$ **B.** $31 \times 40 =$ **C.** $96 \times 30 =$

D. $92 \times 300 =$ **E.** $68 \times 200 =$ **F.** $47 \times 90 =$

G. $340 \times 9 =$ **H.** $450 \times 4 =$ **I.** $2900 \times 4 =$

Part 7. Multiplication Practice

Use paper and pencil only to solve the following problems. Show your work on a separate piece of paper. Be prepared to explain how you estimated to see if your answers were reasonable.

A. $450 \times 3 =$ **B.** $454 \times 9 =$

C. $2090 \times 6 =$ **D.** $90 \times 13 =$

E. $95 \times 20 =$ **F.** $82 \times 30 =$

G. $63 \times 22 =$ **H.** $75 \times 75 =$

I. $129 \times 7 =$ **J.** $432 \times 6 =$

Part 8. Metric Measurement

Use kilometers, meters, decimeters, centimeters, and millimeters to label these distances.

A. The height of a room is about 3 _____.

B. The length of a sofa is about 20 _____.

C. The width of a sharp pencil point is about 2 _____.

D. The length of a pencil is about 14 _____.

E. The distance from Chicago to Dallas is about 1500 _____.

Part 9. Money

Luis has 3 dimes, 3 nickels, and 3 quarters. Help him find all the possible amounts he can make using 3 of his coins. Show how you organized your work. Order the amounts from smallest to largest.

Part 10. The Shortcut

Decide if the following are divisible without actually dividing them out. Explain your strategy.

A. Is 5367 divisible by 2?

B. Is 546,890 divisible by 10?

C. Is 11,952 divisible by 6, 3, and 2?

D. Is 74,981 divisible by 9?

E. Is 431,895 divisible by 5 and 10?

F. Give a three-digit number that is divisible by 9.

G. Give a four-digit number that is divisible by 5.

Part 11. Shortcuts for Checking Answers

Use what you know about divisibility rules to check the answers to these problems. If you find an incorrect answer, solve the problem and write the correct answer. Use a separate sheet of paper if you need more room to show your work.

A.
$$156 \times 9 = 1404$$

B.
$$562 \times 9 = 5008$$

C.
$$231 \times 6 = 386$$

D.
$$410 \times 6 = 2460$$

E.
$$1490 \times 3 = 4470$$

F.
$$289 \times 2 = 577$$

UNIT 12

Exploring Fractions

Unit 12: Home Practice

Part 1. *Triangle Flash Cards:* 5s, 10s, 2s, and 3s

Study for the quiz on the multiplication facts for the first two groups—the 5s, 10s, 2s, and 3s. Take home your *Triangle Flash Cards* and your list of facts you need to study.

Here's how to use the flash cards. Ask a family member to choose one flash card at a time. Your partner should cover the corner containing the highest number. This number will be the answer to a multiplication fact. Multiply the two uncovered numbers.

Your teacher will tell you when the quiz will be given on the 5s, 10s, 2s, and 3s. Remember to study only those facts you cannot answer correctly and quickly.

Part 2. Multiplication

1. Solve the following problems in your head. Remember to follow the proper order of operations.

   A. $87 \times 0 = $ _____   **B.** $211 \times 1 = $ _____

   C. $0 \times 1800 = $ _____   **D.** $1 \times 7898 = $ _____

   E. $8 \times 0 + 8 = $ _____   **F.** $7 + 1 \times 10 = $ _____

   G. $16 \times 1 - 7 = $ _____   **H.** $6 \times 1 - 2 \times 0 = $ _____

   I. $20 - 0 \times 7 = $ _____   **J.** $20 \times 7 - 0 = $ _____

2. Explain why all the multiples of 6 are even numbers. First, write all the multiples of 6 in order from 6 to 60.

Name _____ Date _____

Part 3. Fraction Chart

Use the fraction chart you created in Lesson 3 or the Fraction Chart in your *Student Guide* in Lesson 3.

1. Order these fractions from smallest to largest.

   A. $\frac{2}{3}$ $\frac{1}{4}$ $\frac{5}{6}$ $\frac{3}{8}$ $\frac{2}{12}$ _____

   B. $\frac{3}{5}$ $\frac{1}{8}$ $\frac{4}{9}$ $\frac{1}{10}$ $\frac{1}{2}$ _____

2. Find one or two equivalent fractions for each of the following.

   A. $\frac{6}{8}$ B. $\frac{3}{9}$ C. $\frac{2}{3}$ D. $\frac{4}{10}$

   E. $\frac{3}{5}$ F. $\frac{3}{12}$ G. $\frac{6}{12}$ H. $\frac{1}{5}$

   I. $\frac{5}{6}$

3. Name a fraction smaller than each of the following.

   A. $\frac{1}{2}$ B. $\frac{1}{4}$ C. $\frac{3}{5}$ D. $\frac{7}{8}$

4. Name a fraction greater than each of the following. Do not name a fraction equivalent to 1.

   A. $\frac{1}{2}$ B. $\frac{3}{4}$ C. $\frac{1}{6}$ D. $\frac{9}{10}$

Part 4. Fractions and Decimals

Complete this table. The flat equals one whole.

Base-Ten Shorthand	Common Fraction	Decimal Fraction
|| · · · · ·	$\frac{25}{100}$ or $\frac{1}{4}$	0.25
▢ ·		
		0.03
▢▢▢▢▢ ||/||		
	$\frac{3}{10}$	
		4.37
▢▢ · · · ·		
	$4\frac{16}{100}$	
		10.41

Part 5. A Fraction of a Meter

Use your fraction chart from Lesson 3, the Fraction Chart in your *Student Guide*, or a meterstick to help you compare fractions.

1. Name a measurement that is greater than $\frac{1}{2}$ meter but less than $\frac{7}{10}$ of a meter.

2. Which is longer: $\frac{7}{10}$ of a meter or 50 centimeters?

3. Name a measurement that is a little less than $\frac{3}{10}$ of a meter.

4. Name a fraction of a meter that is longer than $\frac{5}{100}$ of a meter and shorter than 0.2 meter.

5. Name a fraction that is less than $\frac{7}{10}$ but more than $\frac{1}{5}$.

6. Name a measurement that is longer than 1.54 meters but shorter than $1\frac{9}{10}$ meters.

7. Name a measurement that is more than three times as long as $\frac{1}{2}$ of a meter.

Part 6. Arithmetic Review

1. Solve the following problems using paper and pencil only. Estimate to make sure your answers are reasonable.

   A. $231 \times 4 =$ **B.** $409 \times 5 =$ **C.** $6283 \times 4 =$ **D.** $570 \times 5 =$

   E. $46 \times 92 =$ **F.** $27 \times 44 =$ **G.** $73 \times 40 =$ **H.** $83 \times 50 =$

   I. $1092 + 378 =$ **J.** $3807 - 928 =$ **K.** $3450 + 4750 =$ **L.** $8367 - 538 =$

2. Solve the following problems.

   A. John's uncle is taking a test to be a cashier at a grocery store. He must show all the ways to give 59¢ in change using exactly 10 coins. Show at least two ways to do this.

   B. The questions on a math test are each worth a certain number of points. Use the table to find the total points if the four questions are answered correctly.

Question	Points Possible
A	$2\frac{1}{2}$
B	$3\frac{1}{4}$
C	5
D	$4\frac{1}{4}$

Making Fraction Strips

Carefully cut each paper strip from this page. These strips will be used to fold fractions.
Keep your finished strips in an envelope.

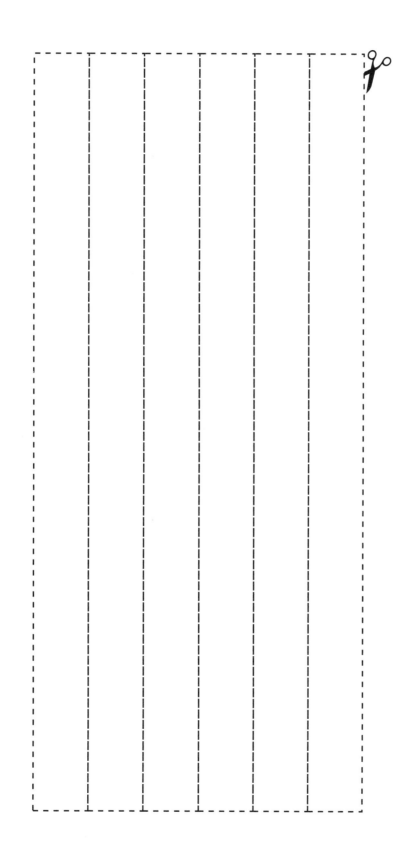

Making Fraction Strips

Carefully cut each paper strip from this page. These strips will be used to fold fractions.
Keep your finished strips in an envelope.

Standard Frabble Cards

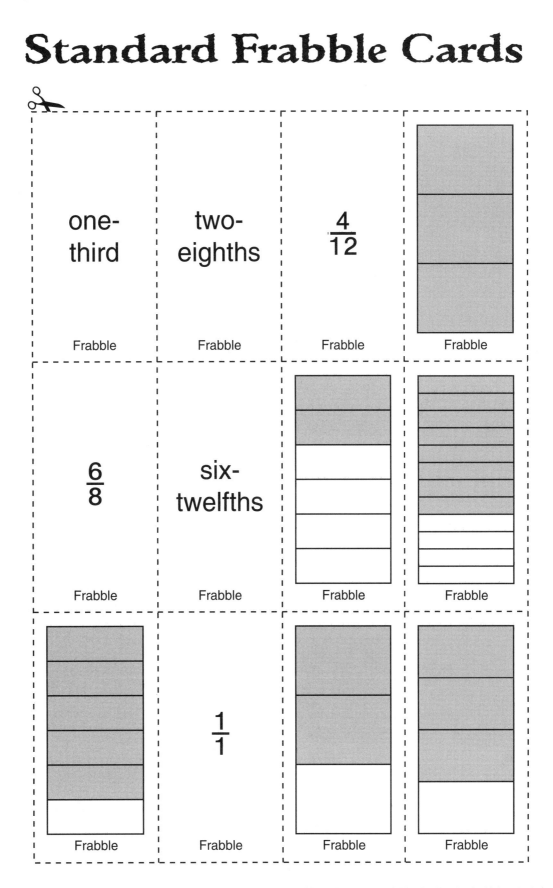

one-third	two-eighths	$\frac{4}{12}$	
Frabble	Frabble	Frabble	Frabble
$\frac{6}{8}$	six-twelfths		
Frabble	Frabble	Frabble	Frabble
	$\frac{1}{1}$		
Frabble	Frabble	Frabble	Frabble

one-sixth	$\dfrac{9}{12}$	ten-twelfths	$\dfrac{1}{2}$ START
Frabble	Frabble	Frabble	Frabble
	$\dfrac{1}{4}$	four-sixths	
Frabble	Frabble	Frabble	Frabble
	$\dfrac{1}{8}$	two-twelfths	one
Frabble	Frabble	Frabble	Frabble

Wild Cards for Frabble

Challenge Frabble Cards

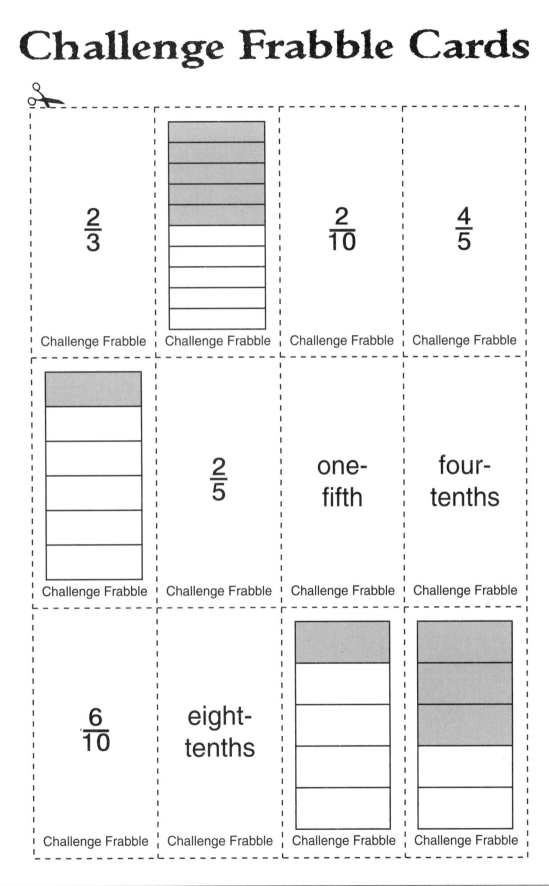

$\frac{2}{3}$		$\frac{2}{10}$	$\frac{4}{5}$
Challenge Frabble	Challenge Frabble	Challenge Frabble	Challenge Frabble
	$\frac{2}{5}$	one-fifth	four-tenths
Challenge Frabble	Challenge Frabble	Challenge Frabble	Challenge Frabble
$\frac{6}{10}$	eight-tenths		
Challenge Frabble	Challenge Frabble	Challenge Frabble	Challenge Frabble

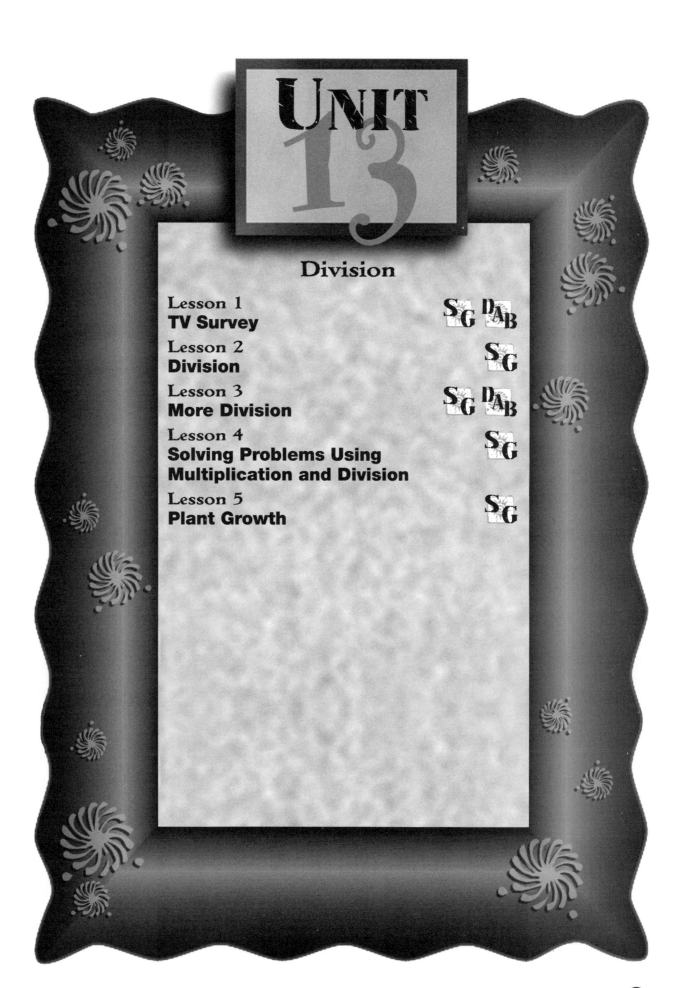

UNIT 13

Division

Unit 13: Home Practice

Part 1. *Triangle Flash Cards:* The 9s, Square Numbers, and the Last Six Facts

Study for the quiz on the multiplication facts for the 9s, square numbers (2×2, 3×3, etc.), and the last six facts (4×6, 4×7, 4×8, 6×7, 6×8, and 7×8). Take home your *Triangle Flash Cards* and your list of facts you need to study.

Here's how to use the flash cards. Ask a family member to choose one flash card at a time. Your partner should cover the corner containing the highest number. This number will be the answer to a multiplication fact. Multiply the two uncovered numbers.

Your teacher will tell you when the quiz on these facts will be. Remember to study only those facts you cannot answer correctly and quickly.

Part 2. Working with Remainders
Show how you solved each of the following problems. Explain how any remainders affected your answer.

1. Nine cereal boxes fit into one crate. How many crates are needed for 30 boxes of cereal?

2. Twenty-one children try out for two teams. The children decided that anyone who is not selected will be an umpire. There are three umpires. How many children are on each of the two teams?

3. Mrs. Roberts collects a total of $273 from the 90 students who are going on a field trip. Each student brings in $3. Is the total of $273 the correct amount? Explain.

4. Irma wants to read her 453-page book in 9 days. If she reads about the same number of pages each night, how many should she read a night?

Part 3. Multiplication and Division Practice

Use paper and pencil to solve the following problems. If you need more work space to show your work, you may use a separate sheet of paper. Estimate to make sure your answers are reasonable.

A. $267 \times 7 =$ **B.** $78 \times 23 =$ **C.** $30 \times 58 =$ **D.** $73 \times 400 =$

E. $6094 \times 8 =$ **F.** $269 \div 3 =$ **G.** $379 \div 2 =$ **H.** $467 \div 8 =$

I. $3708 \div 6 =$ **J.** $4587 \div 7 =$ **K.** $6675 \div 5 =$ **L.** $855 \div 9 =$

Part 4. Create a Fraction

1. You may use your own fraction chart or the Fraction Chart in the *Student Guide* in Unit 12, Lesson 3 to help you solve these problems.

   A. Write a fraction that is bigger than $\frac{1}{2}$, but smaller than $\frac{3}{4}$. _____

   B. Write a fraction that is a little bit smaller than $\frac{1}{8}$. _____

   C. Write a fraction that is double $\frac{1}{10}$. _____

   D. Write a fraction that is much bigger than $\frac{1}{3}$, but smaller than $\frac{11}{12}$. _____

Part 5. Prizes

The children's hospital plans to spend $100 on 100 stuffed animals for prizes for a raffle. The tiny ones are 10¢, the medium ones cost $2.00, and the large ones cost $5. They will buy some of each kind. (There are many solutions to this problem. If you need more work space, use a separate sheet of paper.)

1. How many of each can they buy?

2. What strategy or strategies did you use to solve this problem?

3. How did you check your answer to see if it was correct?

4. Show another solution to this problem.

Part 6. Solving Problems
Choose appropriate methods and tools to solve the following problems.
Explain how you solved each problem.

1. Over a four-day period, Frank watched 325 minutes of TV. He watched 60 minutes of TV on Monday, 45 minutes on Tuesday, and 75 minutes on Wednesday.

   A. How many minutes did he watch on Thursday?

   B. About how many hours of TV did he watch on Thursday?

2. On Monday, Shannon watched 4 times the amount of TV as Luis. If Shannon watched TV for 180 minutes, how many minutes did Luis watch?

3. Mrs. Dewey watched 45 minutes of TV on Monday, 35 minutes on Tuesday, and 1 hour on Wednesday. She did not watch TV on Thursday. What is the mean number of minutes of TV watched by Mrs. Dewey over the four-day period?

4. Mrs. Randall's class watched a total of 5430 minutes of TV over a four-day period.

   A. Is this more or less than 50 hours of TV? How do you know?

   B. Is this more or less than 100 hours of TV? How do you know?

   C. Estimate the number of hours of TV the students in Mrs. Randall's class watched.

5. Mrs. Randall's class decided to change their TV habits. In a follow-up survey, the class watched a total of 3780 minutes of TV in four days.

   A. Did they watch more or less TV? How many minutes more or less?

   B. As a class, how many hours of TV did they cut out of their daily routine?

Daily TV Time

Program	Starting Time	Ending Time	Minutes Watched

Day ____

Total TV Minutes _____

Program	Starting Time	Ending Time	Minutes Watched

Day ____

Total TV Minutes _____

Program	Starting Time	Ending Time	Minutes Watched

Day _____

Total TV Minutes _____

Program	Starting Time	Ending Time	Minutes Watched

Day _____

Average Number of Minutes
per Day _____

Average Number of Hours
per Day _____

Total TV Minutes _____

Total Number of
Minutes for 4 Days _____

How Much TV Do We Watch?

Average TV Time per Day (in hours)	Number of Students	
	Tallies	Total

Division Dot-to-Dot

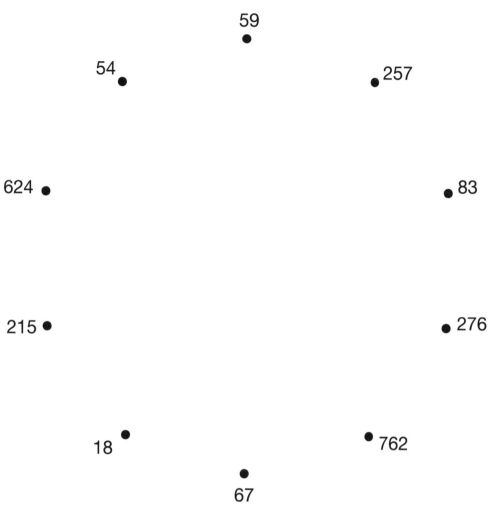

59

54

257

624

83

215

276

18

762

67

Complete each division problem using the forgiving method. Show your work. Locate each answer. Connect the dots in order to form a picture. Use a ruler or a straightedge. Connect your last point with your first point.

1. 472 ÷ 8

2. 828 ÷ 3

3. 162 ÷ 9

4. 324 ÷ 6

5. 332 ÷ 4

6. 603 ÷ 9

7. 1872 ÷ 3

8. 1028 ÷ 4

9. 6858 ÷ 9

10. 1505 ÷ 7

UNIT 14

Chancy Predictions: An Introduction to Probability

Unit 14: Home Practice

Part 1. *Triangle Flash Cards:* All the Facts

Study for the quizzes on all the multiplication facts. Take home your *Triangle Flash Cards* and your list of facts you need to study.

Here's how to use the flash cards. Ask a family member to choose one flash card at a time. Your partner should cover the corner containing the highest number. This number will be the answer to a multiplication fact. Multiply the two uncovered numbers.

Your teacher will tell you when the quiz on the facts will be. Remember to study only those facts you cannot answer correctly and quickly.

Part 2. School Supplies

The school store sells school supplies. The prices for notebooks, pencils, markers, and folders are shown.

spiral notebook	pencils	markers	folders
$ 0.39	4 for $1.00	$2.98	10 for $1.00

Jessie bought: 4 spiral notebooks, 1 set of markers, 8 folders, and 10 pencils.

Jacob bought: 5 spiral notebooks, 1 set of markers, 10 folders, and 5 pencils.

Choose an appropriate method to solve each of the following problems. For each question, you may choose to use paper and pencil, mental math, or a calculator. Be prepared to tell the class how you solved each problem. Use a separate sheet of paper to show your work.

1. Figure out the exact cost of each student's supplies.

2. Whose supplies cost the most?

3. What is the difference in price between the two students' supplies?

4. Choose one of the students' lists and explain how you figured out the total. What was the hardest part of the problem? Why?

Part 3. Multiplication and Division Practice

Use paper and pencil to solve the following problems. If you need more work space to show your work, you may use a separate sheet of paper. Estimate to make sure your answers are reasonable.

A. $945 \times 5 =$ **B.** $312 \times 7 =$ **C.** $99 \times 99 =$ **D.** $6 \times 3274 =$

E. $1342 \times 7 =$ **F.** $1735 \div 5 =$ **G.** $9276 \div 2 =$ **H.** $738 \div 8 =$

I. $9987 \div 6 =$ **J.** $1258 \div 3 =$ **K.** $2232 \div 4 =$ **L.** $9900 \div 9 =$

Part 4. Probability Lines
Use letters to place the following events on the probability line below.

Probability Line

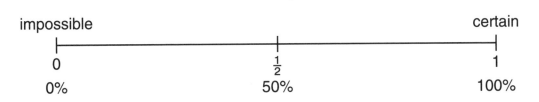

impossible   certain

0 $\frac{1}{2}$ 1

0%   50%   100%

A. Romesh will be the next president of the United States.

B. A spinner is evenly divided into four colors: red, blue, green, and yellow. In one spin, it will land on green.

C. You will travel to Europe next year.

D. Martians will invade the planet this afternoon.

E. A coin flip will land on heads.

F. It will rain tomorrow.

G. There will be seven days in next week.

H. The sidewalk in front of the school is concrete.

I. A dog will have kittens.

J. You will win the lottery today.

Part 5. The Power of 10
Use pencil and paper, not a calculator, for these problems.

A. What patterns do you notice in the problems?

B. What pattern do you think you will see in the answers?

C. 832	D. 832	E. 832	F. 832
$\times 6$	$\times 60$	$\times 600$	$\times 6000$

G. 759	H. 759	I. 759	J. 759
$\times 6$	$\times 60$	$\times 600$	$\times 6000$

K. Arrange your answers from smallest to largest.

Part 6. Movie Library
Solve the following problems using appropriate tools. Explain how you solved each problem.

1. Michael, Shannon, and Jessie decided to create a movie library for their neighborhood. They asked parents and teachers to donate children's videos to create a library. Neighborhood residents could take out a children's movie for free if they brought one in as a trade. In the first week of the drive to collect movies, Michael collected 17 boxes. Each box was filled with 22 movies. How many movies did Michael collect?

2. Shannon collected 11 boxes with 27 movies in each. How many movies did Shannon collect?

3. Jessie and her friends collected 36 boxes with 15 movies in each. How many movies did they collect?

4. Approximately, how many movies did Michael, Shannon, and Jessie collect in all?

5. Jessie's little brother likes to watch the same video over and over. If the video lasts 39 minutes and he watched it 8 times in one week, about how many hours did he spend watching the video?

6. In all, 8 children worked in the movie library on Saturday. If each child helped about 30 neighborhood residents check out one movie, about how many movies were checked out on Saturday?

A Spinner for a
Number Cube

With a partner, use the outline below to create a spinner which gives the same results as a number cube. Use a protractor to help you.

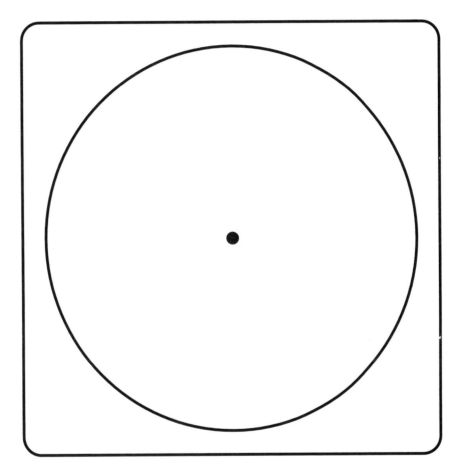

Exploring Spinners 1 and 2

Spinner 1

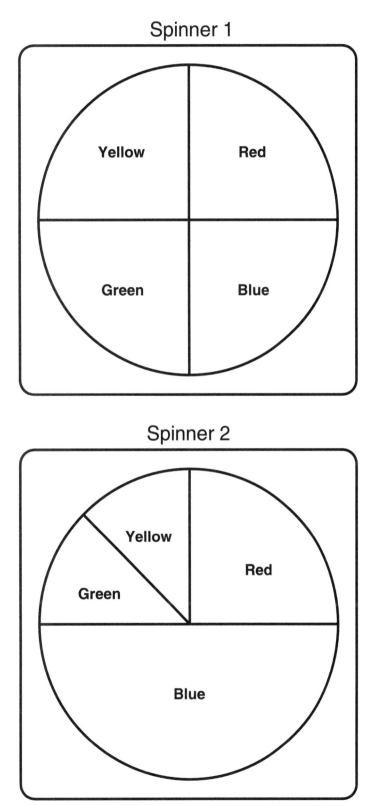

Spinner 2

Blank Spinners

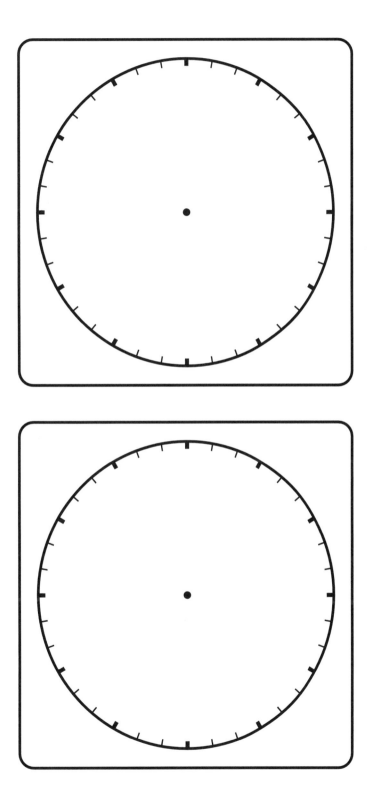

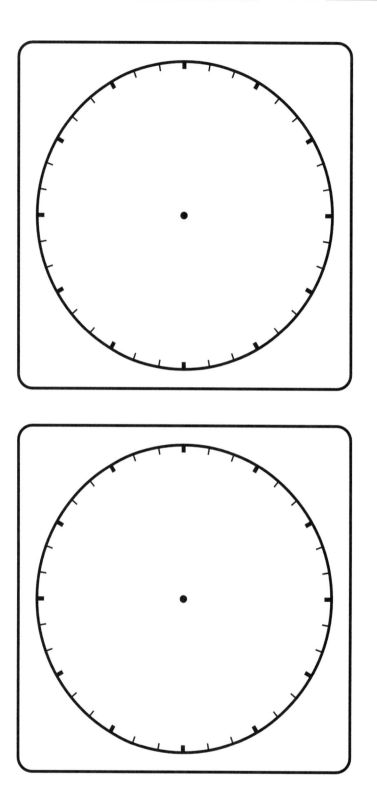

Make Your Own Spinners

UNIT 13

Using Patterns

Unit 15: Home Practice

Part 1. *Triangle Flash Cards:* All the Facts

Study for the two quizzes on all the multiplication facts. Half the facts will be on the first quiz. The other half will be on the second quiz. Take home your *Triangle Flash Cards* and your list of facts you need to study.

Here's how to use the flash cards. Ask a family member to choose one flash card at a time. Your partner should cover the corner containing the highest number. This number will be the answer to a multiplication fact. Multiply the two uncovered numbers.

Your teacher will tell you when the quizzes on the facts will be. Remember to study only those facts you cannot answer correctly and quickly.

Part 2. Mixed-Up Multiplication Tables

1. Complete the tables.

A.

×	3	5	6	8	10
2					
4		20			
7					
8					

B.

×	1	4	6	8	9
3					
6		24			
9					
10					

2. Solve each fact. Then, on a separate sheet of paper, name three other facts that are in the same fact family. For example, the following four facts are in the same fact family: $3 \times 6 = 18$, $6 \times 3 = 18$, $18 \div 3 = 6$, and $18 \div 6 = 3$. (Remember, the square numbers only have two facts in each family.)

A. $8 \times 5 =$ _____

B. $81 \div 9 =$ _____

C. $7 \times 6 =$ _____

D. $9 \times 4 =$ _____

E. $21 \div 7 =$ _____

F. $10 \times 6 =$ _____

G. $4 \times 4 =$ _____

H. $56 \div 7 =$ _____

I. $6 \times 9 =$ _____

Part 3. Practicing the Operations

On a separate sheet of paper, solve the following problems using paper and pencil. Estimate to make sure your answers are reasonable.

A. 546 + 89 = _____   B. 3438 − 723 = _____

C. 2905 + 376 = _____   D. 79 × 5 = _____

E. 2306 × 8 = _____   F. 347 ÷ 5 = _____

G. 62 × 40 = _____   H. 5073 − 782 = _____

I. 9540 ÷ 6 = _____   J. 504 ÷ 9 = _____

K. 1789 + 4532 = _____   L. 6730 − 762 = _____

M. 29 × 44 = _____   N. 4003 ÷ 7 = _____

Part 4. Telling Time

1. What time is it? _____

2. What time will it be in 3 hours? _____

3. What time was it 45 minutes ago? _____

4. What time will it be in $1\frac{1}{2}$ hours? _____

5. What time was it 90 minutes ago? _____

6. Jacob's grandmother is coming to Chicago for a visit. Her plane takes off in Florida at 11:30 A.M. It will take her about 45 minutes to get to the airport. If she wants to arrive at the airport about $1\frac{1}{2}$ hours before take-off, what time should she leave her home?

7. Irma's brother is in high school. He has four 55-minute classes before lunch. If his first class starts at 8:05 and there are 5 minutes between each class, what time is his lunch period? Show how you decided.

Part 5. Solving Problems

Choose an appropriate method to solve each of the following problems. For some questions, you may need to find an exact answer, while for other questions you may only need an estimate. For each question, you may choose to use paper and pencil, mental math, or a calculator. Be prepared to tell the class how you solved each problem.

1. Nila has $585 in her savings account. On her birthday, she deposits $75 she got for birthday gifts. How much money is in her savings account after her birthday?

2. Jackie and her family are taking a 32-mile ferry ride to an island in Lake Michigan. A round-trip ferry ride ticket costs $29 per adult and $15 per child. If 4 adults and 3 children purchase tickets, how much will the ferry ride cost the entire family?

3. John's older brother is in college. His brother and his three roommates want to buy a new stereo that costs $764. If they split the cost of the stereo evenly, how much should each student pitch in?

4. Ming built a house of cards. Before the house came tumbling down, he used 2 full decks of cards. The house also contained all but 15 cards from a third deck. About how many cards were in Ming's house of cards? (A deck of cards has 52 cards.)

5. On vacation, Shannon's family took 3 rolls of 24 pictures and 2 rolls of 36 pictures. How many pictures did the family take in all?

6. Roberto is driving with his family to visit his grandmother. After driving 144 miles from Chicago, the family stops for lunch. They drive 89 more miles and stop for gas. Then, they stop for a soft drink after driving 123 more miles. Roberto's grandma lives 375 miles from Chicago. About how many more miles must they drive before they reach their grandmother's house?

7. If one year is 365 days, how many days old will you be when you are 16 years old?

Part 6. Function Machines

1. Complete each data table using the rules provided.

A.

Input N	Output 8 × N − 4
1	4
3	
5	
7	
9	
11	

B.

Input N	Output 50 − N × 2
2	46
4	
6	
8	
10	
12	

2. Find the rule for each function machine. Then, find the missing numbers in each of the tables.

A.

Input N	Output ____
11	5
15	9
23	17
	27
	53
100	

B.

Input N	Output ____
4	80
5	100
	140
9	180
	200
30	

Good and Bad Experiments

1. Jacob wants to see if the volume of the dirt in the cup affects how tall a plant grows. He does the experiment shown below.

Volume of dirt = 100 cc   Volume of dirt = 300 cc

Is this a good experiment to find out whether the volume of dirt will affect the height of the plant? Explain.

2. Jacob changes his experiment as shown below. Will the experiment now show how the volume of dirt affects plant height? Explain.

Volume of dirt = 100 cc   Volume of dirt = 300 cc

UNIT 16

Assessing Our Learning

Unit 16: Home Practice

Part 1. *Triangle Flash Cards:* All the Facts

Study for the test on all the multiplication facts. Take home your *Triangle Flash Cards* and your list of facts you need to study.

Here's how to use the flash cards. Ask a family member to choose one flash card at a time. Your partner should cover the corner containing the highest number. This number will be the answer to a multiplication fact. Multiply the two uncovered numbers.

Your teacher will tell you when the test on all the facts will be. Remember to study only those facts you cannot answer correctly and quickly.

Part 2. Multiplication Tables

Complete the following tables.

1.

×	7	4	6	2	9
8					
3		12			
5					
1					

2.

×	10	5	6	3	0
8					
4		20			
2					
1					

3. Complete the following. Remember to follow the correct order of operations. Do all multiplications before any addition and subtraction. For example,
$4 + 100 \times 3 =$
$4 + 300 = 304.$

A. $7 \times 100 + 3 =$ _____

B. $6 + 800 \times 7 =$ _____

C. $5000 - 400 \times 5 =$ _____

D. $20 \times 20 + 350 =$ _____

E. $600 \times 80 - 2000 =$ _____

F. $20{,}000 - 18{,}000 \times 1 =$ _____

Part 3. Symmetry
Use the figure below to answer the following questions.

1. Does the figure have turn symmetry? If so, what kind? (*Hint:* Does it have $\frac{1}{2}$-turn symmetry, $\frac{1}{3}$-turn symmetry, $\frac{1}{4}$-turn symmetry, or $\frac{1}{6}$-turn symmetry?)

2. Does this shape have line symmetry? If so, draw in the lines of symmetry.

Part 4. Fractions and Decimals
Complete the following table. ▢ **is one whole.**

Base-Ten Shorthand	Decimal	Fraction
▢ ▢ ▢ / /		
	4.03	
	10.41	
▢ ▢ ▢ ▢ / ...		

Part 5. School Days

Use the data below to answer the following questions.

Country	Average Number of School Days
US	180
Canada	186
England	192
Japan	243

1. About how many days will a child in England have gone to school when he or she has attended school for 5 years?

2. Yoshi lives in Japan. He has gone to school for 5 years. Jackie lives in the United States. She has gone to school for 5 years. About how many more days has Yoshi gone to school than Jackie?

3. About how many days does a student in the USA have to go to school to graduate from high school? (To graduate from high school, most students must go to school for 13 years.)

Part 6. Multiplication and Division Practice

Solve the following problems using paper and pencil or mental math. Show your work on a separate sheet of paper. Estimate to make sure your answers are reasonable.

A. $64 \times 77 =$ _____

B. $23 \times 48 =$ _____

C. $70 \times 56 =$ _____

D. $3540 \times 4 =$ _____

E. $875 \div 7 =$ _____

F. $5960 \div 6 =$ _____

G. $3000 \div 50 =$ _____

H. $7812 \div 9 =$ _____

Part 7. Function Machines

1. Complete the following tables.

A.

Input N	Output 3 + 2 × N
1	5
2	
3	
4	
5	
6	
7	
8	

B.

Input N	Output 5 × N
1	5
2	
3	
4	
5	
6	
7	
8	

2. Nicholas predicted that the outputs would be the same for both data tables. Are they the same? Why or why not?

Part 8. Downhill Racer
You will need a piece of *Centimeter Graph Paper.*

1. John rolled a car down a ramp at three different heights. At each height he took three trials. The data here shows the average distance (in cm) the car rolled at each height. Graph the data on *Centimeter Graph Paper.*

2. If the points form a line, fit a line to the points.

3. Predict the distance the car will roll if the height of the ramp is 30 cm.

Downhill Racer

Height (in cm)	Average Distance (in cm)
8	100
16	198
24	305

Experiment Review Chart

Directions:
- Write the names of the experiments completed this year in the first row of the table.
- Complete each column with information from each lab.

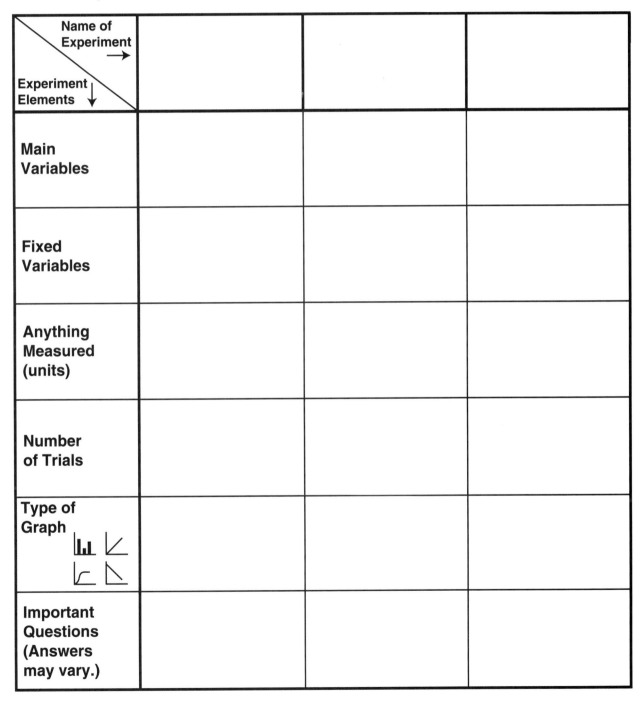

Name of Experiment → Experiment Elements ↓			
Main Variables			
Fixed Variables			
Anything Measured (units)			
Number of Trials			
Type of Graph			
Important Questions (Answers may vary.)			

Name _____ Date _____

Directions:

- Write the names of the experiments completed this year in the first row of the table.
- Complete each column with information from each lab.

Name of Experiment → Experiment Elements ↓			
Main Variables			
Fixed Variables			
Anything Measured (units)			
Number of Trials			
Type of Graph			
Important Questions (Answers may vary.)			